NEXUS NETWORK JOURNAL   Architecture and Mathematics

Aims and Scope

Founded in 1999, the *Nexus Network Journal* (NNJ) is a peer-reviewed journal for researchers, professionals and students engaged in the study of the application of mathematical principles to architectural design. Its goal is to present the broadest possible consideration of all aspects of the relationships between architecture and mathematics, including landscape architecture and urban design.

Editorial Office

Editor-in-Chief
Kim Williams
Via Cavour, 8
10123 Turin (Torino), Italy
E-mail: kwb@kimwilliamsbooks.com

Contributing Editors
The Geometer's Angle
Rachel Fletcher
113 Division St.
Great Barrington, MA 01230, USA
E-mail: rfletch@bcn.net

Book Reviews
Sylvie Duvernoy
Via Benozzo Gozzoli, 26
50124 Firenze, Italy
E-mail: syld@kimwilliamsbooks.com

Corresponding Editors
Alessandra Capanna
Via della Bufalotta 67
00139 Roma Italy
E-mail: alessandra.capanna@uniroma1.it

Tomás García Salgado
Palacio de Versalles # 200
Col. Lomas Reforma, c.p. 11930
México D.F., Mexico
E-mail: tgsalgado@perspectivegeometry.com

Robert Kirkbride
studio 'patafisico
12 West 29 #2
New York, NY 10001, USA
E-mail: kirkbrir@newschool.edu

Andrew I-Kang Li
The Chinese University of Hong Kong
Shatin, N.T.
Hong Kong S.A.R. China
E-mail: andrewili@cuhk.edu.hk

Michael J. Ostwald
School of Architecture and Built Environment
Faculty of Engineering and Built Environment
University of Newcastle
New South Wales, Australia 2308
E-mail: michael.ostwald@newcastle.edu.au

Vera Spinadel
The Mathematics & Design Association
José M. Paz 1131 - Florida (1602)
Buenos Aires, Argentina
E-mail: vspinade@fibertel.com.ar

Igor Verner
The Department of Education in Technology and Science
Technion - Israel Institute of Technology
Haifa 32000, Israel
E-mail: ttrigor@techunix.technion.ac.il

Stephen R. Wassell
Department of Mathematical Sciences
Sweet Briar College
Sweet Briar, Virginia 24595, USA
E-mail: wassell@sbc.edu

João Pedro Xavier
Faculdade de Arquitectura da Universidade do Porto
Rua do Gólgota, 215
4150-755 Porto, Portugal
E-mail: jpx@arq.up.pt

Subscription Information

Nexus Network Journal – Architecture and Mathematics is published by Kim Williams Books, Torino, in one volume per year consisting of two issues, and distributed by Birkhäuser Verlag, Basel. It is also available in electronic form. For further information please visit: http://www.birkhauser.ch

Subscription orders should be addressed to:
Birkhäuser Verlag AG
c/o Springer Customer Service Center GmbH
Customer Service Journals
Haberstrasse, 7
D-69126 Heidelberg, Germany
Phone: ++49 6221 345 43 03, Fax: ++49 6221 345 42 29
email: subscriptions@birkhauser.ch
Business hours:
Monday to Friday 8.00 a.m. to 8.00 p.m. local time

or through your bookseller or subscription agency.

Surface Airmail Lifted (SAL): For subscribers in Japan, India, Australia, and New Zealand, price lists indicating special carriage charges are available.
Airmail delivery to all other countries is available upon request.
Delivery (North America): All journals are airspeeded to subscribers in the USA and Canada.

T0224455

Nexus Network Journal

GUARINO GUARINI:
OPEN QUESTIONS, POSSIBLE SOLUTIONS

VOLUME 11, NUMBER 3
Winter 2009

KIM WILLIAMS BOOKS

Nexus Network Journal
Vol. 11
No. 3
Pp. 329-494
ISSN 1590-5896

CONTENTS

Like that of Palladio in the sixteenth century, the work of Guarino Guarini (1624-1683) in the seventeenth century, both written and constructed, embodies the nexus of architecture and mathematics better than that of anyone else in the Baroque age. Guarini was both a mathematician and an architect, but he was also well versed in their sister arts, including philosophy, stereotomy, geodesy, gnomonics, astronomy and more. Guarini is as rich as he is illusive: he left behind a corpus of ponderous works written in Latin, almost unstudied today, and of his many buildings, only a handful survive, the others victims of disasters both natural (S. Maria Annunziata in Messina destroyed by an earthquake) and manmade (the Chapel of the Holy Shroud destroyed by fire following restoration in 1997). But the fire that destroyed the Chapel of the Holy Shroud offered a unique opportunity to study how Guarini conceived and constructed his masterpieces. This argument was first addressed in these pages in an interview with Mirella Macera (Superintendent for architectural, landscapes, and historical monuments of Piedmont), Fernando Delmastro and Paolo Napoli (see *NNJ* vol. 6 no. 2, 2004). In 2006 a conference entitled "Guarino Guarini: Open Questions, Possible Solutions", dedicated to the Chapel of the Holy Shroud in Turin and its designer, Guarini, was organized by myself and Franco Pastrone of the Department of Mathematics of the University of Turin, and sponsored by the Archivio di Stato and the Direzione per i beni culturali e paesaggistici del Piemonte (see the conference report by Sylvie Duvernoy in *NNJ* vol. 9, no. 1, 2007). The papers in this present issue of the NNJ grew out of that meeting, and will help increase our understanding of Guarini, his scientific and architectural works, and the seventeenth-century context in which he worked.

Several of the papers here provide ample evidence that Guarini worked on a cosmic scale, that he saw both architecture and mathematics not as individual disciplines but rather as interconnected expressions of a universal order. Patricia Radelet-de Grave's paper "Guarini et la structure de l'Univers" focuses on his works in astronomy, especially the 1683 *Coelestis mathematicae*. Here Guarini is placed in the context of other scientists, including Copernicus, Galileo, Kepler, Brahe and Stevin. In "Guarino Guarini and his Grand Philosophy of Sapientia and Mathematics", James McQuillan sets out to reconstruct Guarini's conception of the cosmic order under neo-Platonic-Aristotelian norms and show how this is reflected in his *Architettura civile*. Clara Silvia Roero provides an overview of Guarini's mathematical works, especially his *Euclides adauctus et methodicus mathematicaque universalis*, printed in Torino in 1671.

Guarini, of course, like Leonardo, was a man of his times and did not work in isolation. Paolo Freguglia, in "Reflections on the Relationship between Perspective and Geometry in the Sixteenth and Seventeenth Centuries", provides an overview of the legacy of the 1500s inherited by Guarini regarding the representation of space and techniques of perspective, and shows how these paved the way for future developments in the work of Desargues and Pascal. Michele Sbacchi undertakes a similarly broad examination of the notion of projection in an architectural context. In "Projective Architecture", he places Guarini's ideas of projection in the context of those of Gregorius Saint Vincent, Alberti, Desargues and de l'Orme, among others.

Guarini's built work is also the object of studies presented here. Paolo Napoli, the engineer engaged in the project of rehabilitating the Chapel of the Holy Shroud in the aftermath of the 1997 fire, provides "A Structural Description of the Chapel of the Holy

Shroud in Torino", containing the very latest findings brought to light regarding the damaged chapel. In "Unfolding San Lorenzo" Ntovros Vasileios investigates the architecture of Guarini's Real Chiesa di San Lorenzo in light of the philosophical notion of "fold" introduced by Gilles Deleuze. Ugo Quarello provides a fascinating insight into Guarini's structural prowess in his paper, "The Unpublished Working Drawings for the Nineteenth-Century Restoration of the Double Structure of the Real Chiesa di San Lorenzo in Torino", by examining a part of the architecture that we don't see, that is, the complex masonry structural system behind the Baroque interior. Pietro Totaro looks at "A Neglected Harbinger of the Triple-Storey Façade of Guarini's SS. Annunziata in Messina", tracing the development of one of the most original inventions of the Baroque period in terms of form and proportion, which first appeared in the work of Giacomo Del Duca and Guarini before spreading to the rest of Italy and Europe.

This issue is completed by two book reviews. *Guarino Guarini*, a colossal collection of essays edited by Giuseppe Dardanello, Susan Klaiber and Henry Millon is bound to become a major point of reference in Guarini studies. In *The Architecture of the New Baroque: A Comparative Study of the Historic and the New Baroque Movements in Architecture*, long-time *NNJ* collaborator Michael Ostwald examines claims that what is called the New Baroque really shares a common basic with historical Baroque.

Research into Guarini is by no means exhausted, and I believe that with this issue the *NNJ* makes a significant contribution.

Kim Williams

Paolo Freguglia

Dipartimento di Matematica
Pura e Applicata
Università di L'Aquila ITALY
paolo.freguglia@technet.it

Keywords : Guarino Guarini,
perspective, geometry,
stereotomy, Gérard Desargues,
Blaise Pascal

Research

Reflections on the Relationship between Perspective and Geometry in the Sixteenth and Seventeenth Centuries

Abstract. Paolo Freguglia examines the relationship between perspective and geometry before Guarini, and more precisely, in the 1500s. The representation of space, which in the pre-Classic age was substantially conceptual and sometimes ideographic, was gradually organised so that it became optical representation, and finally arrived at being able to give a sense of three-dimensions. The techniques of perspective were presented not only as practical rules for drawing in a given manner, in conformation with how observed reality appears to the eye, but were also described according to their geometric underpinnings. Thus were introduced new points of departure for considerations on geometry, as made evident by the work of Desargues and Pascal.

1 Introduction

In this paper I would like to shed some light on the relationship of perspective and geometry before Guarini, and more precisely, in the 1500s. It should be recalled that although Guarini was not a professional mathematician, he was undoubtedly interested in mathematics. In Torino in 1671 Guarini published his *Euclides adauctus et methodicus mathematicaeque universalis*, dedicated to Carlo Emanuele II di Savoia, a work of over 700 pages comprising an introduction and thirty-five books (or chapters). Eclectic in scope, it was mainly intended for teaching purposes. In addition to an analysis of the books of Euclid, there are also explanations of how to calculate with logarithms, how to solve plane and spherical triangles, and classic curves such as conics, quadratics, etc. In the twenty-sixth book Guarini addresses projective geometry, which he had presumably learned and studied during the period from 1662 to 1666 when he was in Paris. It is certain however that he knew perspective techniques (which are themselves different from the geometrical techniques proposed by Desargues and Pascal), which he must have learned in his formative years. We cannot know how much and in what way Guarini was able to connect the theory of perspective to the "projective geometry" of Desargues and Pascal. A marked sensitivity for geometry and, in broader terms, the importance of geometry for architecture, is also found in Guarini's *Architettura civile*, published posthumously in 1737.

The representation of space, which in the pre-Classic age was substantially conceptual and sometimes ideographic, was gradually organised so that it became optical representation, and finally arrived at being able to give a sense of three-dimensionality. The completion of this technical objective of representation, which took the study of optics into account as well, first occurred during the fifteenth century and was consolidated during the sixteenth, thanks to a large number of treatises written by the artists and architects as well as mathematicians. The techniques of perspective were presented not only as practical rules for drawing in a given manner, in conformation with how observed reality appears to the eye, but were also described according to their geometric underpinnings. Thus were introduced new points of departure for considerations on geometry. In the first half of the 1600s this aspect was made particularly evident by the works of Gérard Desargues,[1] who was an architect, and by Blaise Pascal,[2] a mathematician. The geometric results that they presented effectively signaled the birth of projective geometry. "Ideal" elements similar to what we today call "point at infinity" and "line at infinity" were introduced; thus what we have at a conceptual level is an amplification of Euclidean space. The theory of proportions also played a determinant role in these developments. This made it possible to obtain geometric properties relevant to projection, although the context of reference always remains Euclidean. Investigations of the theory of the conics, which was already the object of attentive study in the 1500s, were tied to perspective as well. In this period the conic sections were involved[3] not only in perspective (that is, as it was called then, *perspectiva*), but also regarding the so-called burning glasses (mirrors shaped in order to produce combustion) and in "hour lines" (for the construction of sundials). Such topics, which regard the possible applications of the notions inherent in the conic sections, were studied by mathematicians such as Giovan Battista Benedetti, Christophorus Clavius and Francesco Maurolico. To Maurolico (in 1657), Federico Commandino (in 1655) and Giovan Battista Memo (in 1537) we owe translations of the work of Apollonius, which was a review of the theory of conics. It should also be noted that the conics were also examined from a theoretical point of view by Giorgio Valla (1477-1500) and Johann Werner (1468-1522).

2 Treatises on perspective in the Renaissance

Because of their connection with the great developments in figurative art, treatises on perspective flourished during the Renaissance. The representation of objects belonging to a three-dimensional space on a two-dimensional surface is one of the oldest problems of both figurative art and of astronomy. With regard to Antiquity, theatrical creations on one hand and the construction of planispheres on the other offer two highly significant examples: think of Ptolemy's *Planisferio*, which appeared in Latin in 1507.

The theory of perspective construction should be distinguished from the theory of natural vision, which can be traced back to technical studies in Euclid's *Ottica* and *Catottrica*, that is, we have to distinguish between what was called *perspectiva communis* and *perspectiva artificialis* or *perspectiva pingendi*. The rules regarding techniques of painting, studied and applied by painters, and during the design phase, by sculptors and architects of the Renaissance, take into account, in addition to Euclidean geometry, the so-called "practical geometry", manuals[4] which had been developed since the Middle Ages. Between the fifteenth and the sixteenth centuries we therefore find a series of works that testify to the great interest, not only applicative for speculative as well, with which subjects related perspective were studied with the spirit of the age. Here we give a list, in chronological order, of the most significant of these works in order to provide some idea of what constituted the scientific output in this direction during the Renaissance. As can be seen, some of these works are "commentaries" on works of antiquity:

L. B. Alberti, *Della pictura* (1435, printed in 1511)
Piero della Francesca, *De prospectiva pingendi* (1482, 1487)
Leonardo da Vinci, *Trattato della pittura* (printed in 1651)
J. Pélerin, *De artificiali perspectiva* (1505)
A. Dürer, *Istitutionum geometricorum libri* (1525, 1533)
S. Serlio, *Prospettiva* (1545)
F. Commandino, *In Planisphaerium Ptolomaei Commentarius* (1558)
J. Cousin, *Livre de Perspective* (1560)

D. Barbaro, *La pratica della Perspectiva* (1569)

E. Danti, *La prospettiva di Euclide* (1573)

G. B. Benedetti, *Diversarum speculationum* [...] *liber* (1585)

G. U. Del Monte, *Perspectiva* (1600)

S. Stevin, *Perspectiva* (1605).

The majority of the authors cited above had a special sensitivity to and knowledge of mathematics, and they examined the theme of perspective with a mathematical bent and zeal. Among these, and certainly one of the most important of them was Piero della Francesca (ca. 1410-1492) from Borgo Sansepolcro, master artist, whose works profoundly influenced the history of art. We are not interested in the aesthetic aspects of Piero's work here, but rather in the theoretical underpinnings of perspective as found in his *De prospectiva pingendi*. For Piero perspective is necessary for painting, in the sense that it isn't possible to draw correctly without it. It is almost like a syntactic scheme: its violation means that the pictorial "discourse" is ill expressed; or again, insisting with this analogy, when the rules of counterpoint are violated in a traditional tonal musical composition the result is discordant. It is thus an indispensable technique for the artist. Piero writes in the first book:

> Painting consists in three principle parts, which we say are drawing, commensuration, and colouring. By drawing we mean profiles and outlines that contain the object. By commensuration we mean profiles and outlines that are placed proportionally in their locations. By colouring we mean to apply the colours as they are in the objects they show, light or dark according to the changing lights. Of the three parts I intend to treat only commensuration, which we call perspective... .[5]

In his turn Leonardo da Vinci wrote:

> Perspective is the demonstrative proof that is confirmed by experience that all things communicate their likenesses to the eye by pyramidal lines; and that the pyramids of bodies of equal magnitudes form a larger or smaller angle according to the kind of distance that there is between one and the other... .[6]

This kind of statement is quite similar to those of other artists, such as, for example, Leon Battista Alberti, who writes in Book I of *Della pittura* regarding to the use of the pyramid in optical perspective:

> The pyramid is a figure of a body from whose base straight lines are drawn upward, terminating in a single point. The base of this pyramid is a plane which is seen. The sides of the pyramid are those rays which I have called extrinsic. The cuspid, that is the point of the pyramid, is located within the eye where the angle of the quantity is. Up to this point we have talked of the extrinsic rays of which this pyramid is constructed. It seems to me that we have demonstrated the varied effects of greater and lesser distances from the eye to the thing seen.[7]

As though he were stating postulates, Leonardo wrote that:

> The eye only sees by means of pyramids;[8]

> The path of the pyramidal lines caused by the objects and terminating in the eye are necessarily straight lines;[9]

> The resemblance of the figures and colour of each body are transferred from one to the other by means of pyramids.[10]

He then gives a proof:

The small object from nearby and the large [object] from far away, being seen in equal angles, will appear of equal size.[11]

Leonardo also recorded his thoughts on some geometric-perspective definitions. For example, regarding the point, he follows the geometers in saying "it is that which cannot be divided into any parts" but the point is also "that which, located in the eye, receives all the points of the pyramids".

Let's return to the work of Piero della Francesca.[12] The treatment develops along the pattern of mathematical treatises, although not always with the same kind of rigour. However, as a preliminary we have to underline the fact that perspective in not projective geometry; rather, it constitutes a "Euclidean model" of a technical apparatus that serves to represent – to "see" – a three-dimensional reality in two dimensions. We might say paradoxically that a perspective representation constitutes a Euclidean model of the projection. This can be best clarified with an example. As we know, two straight lines in the Euclidean plane remain as such in a representation; let's say that they have the same direction, and let's add conceptually that they meet in a point at infinity, but that this point does not belong to the Euclidean plane and therefore we can't represent it with any point on this plane. If instead we do choose a point (the vanishing point) in the Euclidean plane and make it so that straight lines that we hold to be and conceive of as parallel converge at that point, then the point is the conventional-perspective representation in the Euclidean plane of a point at infinity.

We can see that graphic rules and equalities between relations form the basis for perspective drawing. The latter, that is, equalities between relations, is based on the theory of proportions.

Piero's work was given a great deal of consideration by authors of perspective in the 1500s. We need only mention the treatise of Daniele Barbaro, who in spite of the fact that he gives just a passing mention to "Pietro dal Borgo S. Stefano" in the 'Proemio', draws on Piero's work a great deal, as has been often pointed out, in the rest of the treatise.

Going back to what was said earlier concerning the relationships between perspective and projective geometry, it is worthwhile to underline the fact that the vanishing points are representations in the Euclidean plane of points at infinity. The perspective plane can thus be seen as the Euclidean representation of the projective plane whose line at infinity, understood as the set of its points at infinity, is the straight line on which we have represented all of the vanishing points and which can be intuitively equated to the horizon.

The treatises on perspective continued to develop autonomously in the 1700s with technical refinements and mathematical methods that were increasingly defined, as can be seen, for example, in the works of Brook Taylor and J. H. Lambert.[13] Descriptive geometry and projective geometry, which owe a great deal to perspective, were given their definitive treatment by, respectively, Gaspard Monge (*Géométrie Descriptive*, 1798) and J. V. Poncelet (*Traité des propriétés projectives des figures*, 1822). However, as Poncelet himself acknowledged, we owe to Gérard Desargues (1593-1662) and Blaise Pascal (1623-1662) the study of the geometric properties, above all of the conics, which although in a different theoretical context than the one introduced by Poncelet, concern projective geometry and were inspired, as we shall see, by the theme of perspective.

3 Towards projective geometry

In some of the propositions, and more precisely from 129 to 145 of the *Collectiones mathematicae* (ca. 320 A.D.) by Pappus of Alexandria, we find a kind of theory of transversals and notions which regard, for example, what we would today call conditions of alignment between points or the invariance of cross-ratios.[14]

However, it was the work of Desargues that played the crucial role for the themes of projective geometry. Although his results did not enjoy either at the time in which he was developing them nor in the periods that followed the success they deserved, it must be admitted that the notions he set out were of great significance historically and conceptually. They effectively broadened the understanding of fundamental Euclidean notions, gave special consideration to the concept of projective transformation, and shed light on the invariance of such transformations of relations among points, as in the case of the involution among points, an idea explicitly introduced by Desargues. The relevant works of Desargues are the *Première proposition géométrique*, as presented by Bosse in 1648 and the *Brouillon Projet d'une atteinte aux événements des rencontres d'une cône avec un plane et aux événements des contrariétés de entre les actions des puissances ou forces* of 1639. In the first Desargues presents the well-known theorem of homologous triangles.

The 1639 *Brouillon Projet*, although essentially an outline, is characterised by an expository organisation and a concise method of geometric investigation that are particularly novel. In this work (as in *Première proposition*) the Ptolemy-Menelaus theorem plays a crucial role in the proofs of the fundamental results in terms of projective geometry, and Desargues specifically mentions it in the *Brouillon Projet* in connection to the case of the plane. Let's begin by presenting the definitions of Desargues regarding what we call elements at infinity and thus of the notion, also introduced by Desargues, of involution.

> We say that several straight lines are of the same ordinance (*ordonnance des lignes droits*) if they are either all parallel or if they all meet in the same point. In both cases these straight lines tend to a point. ... The point to which the straight lines tend in the two cases seen above is called the "goal" of the straight line of a given ordinance (*but d'une ordonnance de droites*). ... We say that several planes are of the same ordinance (*ordonnance de plans*) if they are either all parallel or they all meet in the same straight line. In both cases these planes tend to a straight line. ... The straight line to which the planes tend in the two cases above is called the "goal" of the planes of a given ordinance (*but d'une ordonnance de plans*).

We can see right away that Desargues highlights the fact that from a projective point of view points and straight lines in infinite case both have the same characteristics as the corresponding entities in finite case; in fact, like the latter, the former can be obtained as intersections (at infinity) of straight lines and planes. Desargues then presents the generation of a sheaf of straight lines (analogous to a sheaf of planes) in this way:

> Supposing an infinite straight line having a fixed point moving along its entire length, we see that in the various positions that it assumes during this motion it gives the various straight lines of a single sheaf whose centre is the fixed point.

This way of generating sheafs brings to mind the notion of direction of motion of the first kind in the sheaf of lines, where the straight line generatrix is able to move in two opposite directions starting from an initial position. In this same vein we also have the concept that if a fixed point, the centre of the sheaf, is in the finite case, then each point of the line that is different from the centre which moves, generates a circumference; while in the infinite case the fixed point generates a line perpendicular to the line that moves. Thus Desargues is able to identify a connection of sorts between the infinite straight line and the circle. He likewise introduces the notion of involution between six points on a straight line considered as a set of points, based on and developed from the theory of proportions. What follows shows how Desargues reasons and how he presents and studies the notion of involution.

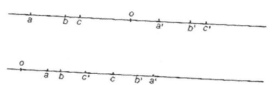

Given a point O on a straight line and the three pairs of segments (called branches) Oa, Oa'; Ob, Ob'; Oc, Oc' (see figure), if the rectangles constructed on the three pairs of segments are equal, that is, if

$$Oa \cdot Oa' = Ob \cdot Ob' = Oc \cdot Oc' \qquad (3.1)$$

then the six points (called nodes) a, a'; b, b'; c, c' are said to be in involution. Let's see now how the involution between three points can be characterised by relations that do not involve point O (called the trunk of the involution). Let's begin with the second part of (3.1), which we rewrite as:

$$Ob' : Oc' = Oc : Ob \qquad (3.2)$$

From (3.2) we obtain:

$$(Ob' + Oc) : Oc = (Oc' + Ob) : Ob$$

that is, summing and then exchanging the means:

$$b'c : bc' = Oc : Ob \qquad (3.3)$$

Comparing (3.2) and (3.3) we have:

$$Ob' : Oc' = Oc : Ob = b'c : bc' \qquad (3.4)$$

Using the same procedure Desargues also obtains:

$$Oc' : Ob = Ob' : Oc = b'c' : bc \qquad (3.5)$$

Multiplying (3.4) by (3.5) member by member, we have

$$Ob'/Ob = (b'c \cdot b'c')/(bc' \cdot bc) \qquad (3.6)$$

Starting from Ob' : Oa'=Oa : Ob=b'a : a'b and Oa' : Ob=Ob' : Oa=b'a' : ab, multiplying member by member we have:

$$Ob'/Ob = (b'a \cdot b'a')/(a'b \cdot ab) \qquad (3.7)$$

Comparing (3.6) and (3.7) we obtain:

$$(b'c \cdot b'c')/(bc' \cdot bc) = (b'a \cdot b'a')/(a'b \cdot ab) \qquad (3.8)$$

In an analogous way Desargues establishes that:

$$(c'b \cdot c'b')/(cb \cdot cb') = (c'a \cdot c'a')/ (ca \cdot ca') \qquad (3.9)$$

$$(a'b \cdot a'b')/(ab \cdot ab') = (a'c \cdot a'c')/(ac \cdot ac') \qquad (3.10)$$

Thus it is established that:

(Theorem): Let there be six points a, a'; b, b'; c, c' on a straight line. Equations (3.8), (3.9), (3.10) characterise each involution of the six points in question.

Desargues thus resorts, as we said, to the Ptolemy-Menelaus theorem and applies it as a lemma to two theorems of signal importance for projective geometry: the so-called theorem of complete quadrilaterals, and the demonstration of the invariance of involution under projection.

4 The conics according to Desargues and Pascal

In the *Brouillon Projet* Desargues applies his results relative to involution and complete quadrilaterals to the study of conic sections. He defines the conics as projections of the circle but, unlike the Greeks, he was able to see general projective properties. According to Desargues the generation of the conics occurs when the following basic conditions are taken into account. If a straight line that has a fixed point moves along a circumference there are two possibilities:

1. the fixed point is in the plane of the circumference and so the straight line with its successive positions describes a sheaf that can be with the centre at infinity or no, depending on whether the fixed point is at an infinite or finite distance;

2. the fixed point is outside the plane of the circle at some finite or infinite distance, so that we obtain a solid that we call cone if the point is at a finite distance, and cylinder if the point is at an infinite distance.

As we can see, a cone and a cylinder are seen in the same way. This is followed by an exposition of the conic sections that is very similar to that by Apollonius. It is worthwhile noting that the cylindrical sections had already been studied between the sixteenth and the seventeenth centuries,[15] but with Desargues we find another way of looking at the subject. Thus the theorem about the quadrilateral inscribed in a conic and cut by a transversal is stated in this way:

Given a quadrilateral inscribed in a conic, a secant line which does not pass through any of the vertices of the quadrilateral cuts the conic in two points that are joined in the involution to which belong the three pairs of points marked on the same straight line by the pair of opposite sides of the quadrilateral (Desargues, *Brouillon Projet*, Theorem 3).

Desargues says that "when well understood, this can be applied on numerous occasions" [Taton 1951: 137 ff]. Thus, as Taton observes, the author observes "the determining role of the projective method that serves him as a guide" [1951: 137 ff] for the investigations proposed in the *Brouillon Projet*.

As is well known, in his *Essay pour les Coniques* of 1639/1640 Blaise Pascal studied some remarkable properties of conics, making reference to the methods of Desargues. Historians of mathematics are almost unanimous regarding Desargues's influence on Pascal, to the point of considering Pascal's results as a follow-up, an amplification and a partial evolution of those of Desargues. Following some definitions, there is Pascal's well-known theorem of a hexagon inscribed in a conic. Pascal precedes this with the following definition: "when several straight lines converge to the same point or are parallel among themselves, the these straight lines are said to be of the same order and the same ordinance...". This is of course thoroughly analogous to what Desargues had said regarding *ordonnance des lignes droites*.

5 Conclusions

I wish to conclude with a few brief historical considerations on the relationships between Descartes and Desargues and Pascal, who knew each other thanks to the centre of attraction and cultural discussion that P. Mersenne created. In a letter from Descartes to Desargues dated 18 June 1639, he refers to the *Bouillon Projet*, which Mersenne had sent to Descartes. In the first part of the letter Descartes disagrees with Desargues's choice of introducing new mathematical terminology (as example of which, and perhaps the only term which remains in use today, is involution) inasmuch as scholars conventionally referred to the classical traditions, which in the case of conics went back to Apollonius; if instead Desargues's intention was write for those who practiced perspective and architecture in a technical sense, then these terms might enjoy a certain success, because they were more closely linked to the French language and to a practical sense. But in this case, the mathematical task that was outlined in the *Projet* would be but little appreciated. Descartes's opinion is typified by the following part of the letter, in which we can clearly see not only that he is suggesting the use arithmetic for doing geometry, but also that he has little appreciation for concise methods. He writes:

> As far as this is concerned, it seems to me that, in order to make your proofs simpler, using arithmetical terms and operations would not be out of place, as I did in my *Géométrie*, because there are many more people who know what a multiplication is and what a composition of three ratios is, etc.

Descartes concludes by saying that he thinks that Desargues's proposal of unifying the notion of points in the finite case and in the infinity case is valid. In a letter dated 12 November 1639 Mersenne had written to Descartes in extremely flattering terms regarding the precocious and brilliant mathematical mind of Pascal. In April 1640 Descartes had a copy of the *Essay* in his hands and considered it to be a unqualified outgrowth of Desargues's concepts.

As I said, Desargues's formulation and methods were not successful in the mathematical world at the time, except for the interest they aroused on the part of Abraham Bosse (1611-1678) and Philippe de la Hire (1640-1718). The attention they deserved would only arrive in the nineteenth century. It was instead the methods of Descartes, undoubtedly due in part to their technical potential, that characterised the developments of geometry and mathematics in the second half of the seventeenth century.

Translated from the Italian by Kim Williams

Notes

1. For a more in-depth analysis of the work of Desargues, cf. [Field 1997]; [Freguglia 1982]; [Poudra 1864] and [Taton 1951].
2. For Pascal's contribution to geometry, and especially his *Essay pour les coniques*, cf. [Taton 1951: 190-194], [Pascal 1904-1914] and [Gehrhardt 1892].
3. Cf. [Clagett 1965-1984], in particular vol. IV, A Supplement on the Medieval Latin Traditions of Conic Sections (1150 – 1566) published in 1980, and [Maierù 1996].
4. For more on Medieval and Renaissance treatises of practical geometry, cf. the various texts edited and commented in the form of notebooks by the Centro Studi per la Matematica Medioevale of the University of Siena.
5. *La pictura contiene in sè tre parti principali, quali diciamo essere disegno, commensuratio et colorare. Disegno intendiamo essere profili et contorni che nella cosa se contiene. Commensuratio diciamo essere essi profili et contorni proportionalmente posti nei luoghi loro.*

Colorare intendiamo dare i colori commo nelle cose se dimostrano, chiari et oscuri secondo che i lumi li devariano. De le tre parti intendo tractare solo de la commesuratione, quale diciamo prospectiva ... . Regarding the work of Piero della Francesco in general, cf. [Panofsky 1993; 1939; 1974]; for *De prospectiva pingendi* cf. [Piero della Francesco 1984].

6. *Prospettiva è ragione dimostrativa, per la quale l'esperienza conferma tutte le cose mandare all'occhio per linee piramidali la lor similitudine; e quali corpi d'equali grandezza faranno maggiore o minore angolo a al lor piramide secondo la varietà de la distanza che sia da l'una a l'altra...* [Leonardo da Vinci, MS. A, fol. 3r].

7. *La piramide sarà figura di un corpo dalla cui base tutte le linee diritte tirate in su terminano ad un solo punto. La base di questa piramide sarà una superficie che si vede. I lati della piramide sono quelli razzi, i quali io chiamo extrinseci. La cuspide, cioè la punta della piramide, sta drento all'occhio quivi dove l'angholo delle quantità. Sino ad qui dicemmo de i razzi extrinseci da i quali dia conceputa la piramide; et parmi pruovato quanto differentii una più che un'altra distantia tra l'occhio et quello che si vegga* [Alberti 1956: bk. I, 46-47].

8. *L'occhio non vede se non per piramide* [Leonardo da Vinci, *Codex Arundel*, fol. 232v].

9. *Il concorso delle linee piramidali causate da li obietti e terminate nell'occhio è necessario essere rette linee* [Leonardo da Vinci, *Codex Arundel*, fol. 9v].

10. *Le similitudini delle figure e colori di ciascun corpo si trasferiscono dall'uno all'altro per piramide* [Leonardo da Vinci, *Codex Arundel*, fol. 232r].

11. *La cosa piccola da presso e la grande da lontano, essendo viste dentro a equali angoli, appariranno d'equale grandezza* [Leonardo da Vinci, *Ms. A*, fol. 8r].

12. For more details, cf. [Boffito 1993: 134-135].

13. Cf. [Freguglia 1995]; [Andersen 1991]; [Lambert 1981].

14. The cross-ratio is the ratio between two simple ratios: given three distinct points A, B, P on an oriented straight line r, it is possible to identify the ratio h = AP/BP (with the condition that AP=−PA and BP=−PB), which, it turns out, depends on neither the orientation of r nor on the unit on measure chosen on r; simple ratio h is conventionally denoted as (ABP). If instead we now consider four points A, B, C, D (in a finite case and at infinity) on r, the cross-ratio b of the four points considered in a given order is:

$$b = (ABCD) = (ABC) : (ABD) = (AC/BC) : (AD/BD).$$

It can be proven that b does not change if two of the elements are interchanged, as long as the other two are also interchanged.

Further, the four points of an oriented straight line r, taken in a given order, form a harmonic set as long as their cross-ratio in that order is equal to −1. To be precise, if A, B, C, D are the four points in given order, then

$$(ABCD) = -1,$$

from which, among other things, it happens that

$$(ABC) = - (ABD).$$

15. Maurolico wrote an important manuscript on cylindrical sections in 1534, entitled *Sereni cylindricorum libelli duo*, in which he reconstructs a treatise by Serenus of Antinouplis (II sec. A.D.) which arrived to the West through the work of Francesco Filelfo (1398-1481) and translated in part by Giorgio Valla; cf. [Tassora 1995].

References

ALBERTI, Leon Battista. 1956. *On Painting*. John R. Spencer, trans. New Haven: Yale University Press.

ANDERSEN, K. 1991. *Brook Taylor's Work on Linear Perspective: A Study of Taylor's Role in the History of Perspective Geometry*. Springer Verlag, New York,.

BOFFITO, M. 1993. *Dentro la geometria. Sui problemi di geometria proiettiva, evoluzione storica e applicazioni*. Genova: Grafic Print.

CLAGETT, M. 1964-1984. *Archimedes in the Middle Ages*, 5 vols. Philadelphia: The American Philosophical Society.

FIELD, J. V. 1997. *The Invention of Infinity*. Oxford: Oxford University Press.

FREGUGLIA, P. 1982. *Fondamenti storici della geometria*. Milan: Feltrinelli.

———. 1995. De la perspective à la géométrie projective: le cas du théorème de Desargues sur les triangles homologiques. In *Entre Mécanique et Architecture*, eds. Patricia Radelet-de Grave and Edoardo Benvenuto. Basel: Birkhäuser.

GERHARDT, C.I. 1892. *Desargues und Pascal über die Kegelschnitte.* Berlin: Akademie der Wissenschaften, 1892.

LAMBERT, J.H. 1981. *Essai sur la perspective* (1752). Trans. J. Peiffer. Paris: Monom ed.

MAIERÙ, L. 1996. *La teoria e l'uso delle coniche nel Cinquecento.* Studi dell'Istituto Gramsci siciliano 17. Caltanisetta and Rome: Salvatore Sciascia Editore.

PANOFSKY, E. 1993. *Perspective as Symbolic Form* (1927). Cambridge, MA: MIT Press.

———. 1939. *Studies in Iconology: Humanistic Themes in the Art of the Renaissance*, Oxford: Oxford University Press.

———. 1974. *Meaning in the Visual Arts* (1955). London: Viking Press.

PASCAL, B. 1904-1914. *Oeuvres complètes publiées suivant l'ordre chronologique.* Eds. L. Brunschvicg, P. Boutroux, F. Gazier. Paris.

PIERO DELLA FRANCESCA. 1984. *De prospectiva pingendi.* Ed. G. Nicco-Fasola. Florence: Le Lettere.

POUDRA, N.G. 1864. *Oeuvres de Desargues réunies et anallysées ... .* 2 vols. Paris.

TASSORA, R. 1995. I Sereni cylindricorum libelli duo di Francesco Maurolico e un trattato sconosciuto sulle sezioni coniche. *Bollettino di Storia delle Scienze Matematiche* XV.

TATON, R. 1951. *L'oeuvre mathématique de G. Desargues.* Paris: Presses Univ. de France.

About the author

Paolo Freguglia is Full Professor at the University of L'Aquila, where he teaches mathematical biology and history of mathematics, which are also his fields of research. He has been visiting professor at several universities outside Italy. He is a chairman and member of the editorial board of several international journals. His scientific production (in all, approximately 150 works between books, essays and articles) principally regards the following areas of research: the construction and analysis of several mathematical models in physics (betatronic motion and geometrical optics) and biology (biomathematical models of neo-Darwinian concepts); the history of mathematics in the sixteenth and seventeenth centuries (regarding Bombelli, Tartaglia, Stevin and in particular François Viète); the history of logic and of mathematics in the nineteenth and early twentieth century, including Peano and his school, and the related themes of the algebra of logic, mathematical logic, the foundations of geometry, and geometric calculus (G. Bellavitis, C. Burali Forti, W.R. Hamilton, etc.).

James McQuillan

23 Cannon Court
Church St.
Cambridge CB4 1ED
GREAT BRITAIN
cosmos231@ymail.com

Keywords: Guarino Guarini,
Baroque architecture,
philosophy of Second
Scholastic and the art of
projection.

Research

Guarino Guarini and his Grand Philosophy of Sapientia and Mathematics

Abstract. This paper proposes that there is a wide and important wealth to the philosophy and art of Guarino Guarini, and one of the keys to these matters lies in the structure and content of his *Architettura civile*, his posthumously-published treatise on architecture. Guarini was an important mathematician in the development of calculus, and his fame is not just that of an architect, perhaps the most learned that Europe has ever thrown up.

Introduction

The purpose of this contribution to Guarinian studies is to establish the ulterior or ultimate meaning of this complex figure of the High Baroque, which has not been fully dealt with previously, leading to even wayward results in the literature. So this will be an exploration of the hermeneutical significance of the polymathic attributes of the Theatine architect, who was also a theologian, an area in which he did not publish, as well as an encyclopaedic writer on philosophy, mathematics and astronomy.

As Joseph Mazzeo put it, the breakdown of hermeneutics in Biblical exegesis, has been followed by a reverse movement – 'the secularization of *hermeneutica sacra* and its amalgamation to *hermeneutica profana* open the way to two opposing tendencies in modern humanistic and literary interpretation, the sacralization of the secular and the secularization of the sacred' [Mazzeo 1978, 24]. Three hundred years ago, such confusion was impossible, and age-old conventions were in place to secure man's place in the world, mainly through a cosmology which was being questioned, of course, but only by an elite. In the time of Milton, everyone was an artist in the sense of participating in thought and enquiry, so that he could call Galileo 'the Tuscan artist'. An encyclopaedic survey of optics and perspective was entitled the 'Great Art of Light and Shadow' (Athanasius Kircher SJ, *Ars magna lucis et umbrae in decem libros digesta*, Rome, 1646). The liberal arts still reigned as the introduction to philosophy both moral and natural (the *trivium*) as well as mathematics (the *quadrivium*). The human mind was a reflection of the cosmic order under neo-Platonic-Aristotelian norms, shared by the great artists and architects of the Baroque, especially the only one who was also a scholar and mathematician: Guarino Guarini (1624-1683). Due to the breakdown in pre-understanding necessary to consider Guarini after his death with the rise of modern science, a fresh effort must be made to restore the conditions to read this Baroque figure as he deserves to be read. Only since 1970 has this Theatine priest, a member of the Clerks Regular who practised as a court architect in Turin where he built the astounding Chapel of the Holy Shroud, been recognised on a par with the architects Bernini and Borromini. The young Guarini studied in Rome where the others were then flourishing, and presumably the novitiate studied these masters in detail.

An attempt has already been made to construct a connection between *Guarini artista* and *Guarini scienziato*: Marcello Fagiolo's 'La *Geosofia* del Guarini' [1970] presumed to explicate Guarini's thought *more geometrica*. As Guarini seems to have made no change to the traditional divisions of the applied arts or sciences, geometry as understood by Fagiolo cannot be given the same basis of equality with philosophy and astronomy that he proposed. Rather for Guarini there are three contiguous worlds in the arts – Philosophy, Mathematics and Medicine – the first and last featuring in his *Placita philosophica* [1665]. The main deficiency of Fagiolo's treatment of Guarini's science is the absence of the transcendental dimension of his Universal Mathematics, and its alignment with metaphysics, made clear in the opening pages of his *Euclides adauctus*. If Guarini reorganised the speculative sciences in terms of Universal Philosophy, he remained content with the traditional model for the arts, and here, too, the ordering followed an order both philosophical and cosmographical – in a word, traditional. For Guarini, Philosophy referred to morals, law and the world of secondary causality, Mathematics is either Macrocosmic or Universal, (transcendental), or Microcosmic, the traditional mixed sciences, and Medicine, the amelioration of our human frailty, as he put it, 'for all arts depend either on Mathematics, or on Philosophy, or on Medicine; because all sciences contemplate either things of similitude, or of proportion, or of convenience'.[1] To further articulate Guarini's *scientia scientiarum*, we must be guided by his portrayal of reality in the *Placita philosophica*, and the consequences of the cosmological model found there. The phenomenon of rotundity in the *Placita philosophica* is matched by the astronomical activities of the celestial bodies, the terrestrial change of generation and corruption is measured by the gnomonical operations of solar horology, both activities further treated mathematically in the Theatine's astronomical works. Finally, the whole structure has its conclusion in *quadratura* – squaring the circle – and transformation, where Guarini's two elements of earth and water suffer further mutation into aberrant forms of the mundane natural world either through monstrous development or through the force of human imagination: this last interpretation is based on clues from the mathematical works of Mario Bettini SJ, (1582-1657, the mathematical writer from Bologna, who produced the most graphic and encyclopaedic presentation of Baroque mathematics, but without algebra), Guarini's use of various methods of projection, many found in Bettini, coupled with a careful reading of his treatise *Architettura civile*.[2] It should be noted that Guarini referred to squaring-the-circle in his *Euclides adauctus*, but did not confirm any method to do so. However, in accord with Fagiolo's presentation, Guarini was insistent in reconciling Universal Mathematics, a pure system as outlined by Euclid and others, with theology, consistent with certain moves to maintain certainty between theology and mathematics developed in Rome in the late sixteenth century.

To sum up at this stage, here is an overview of Guarini's 'grand' philosophy;

- The reconciliation of Plato and Aristotle;
- Scholastic dialectic plus *experientia* or experiment;
- Universal Mathematics linked to light and the Triple Worlds of Early Christian cosmology (the planetary world, the mundane world, and the underworld);
- Hierarchical understanding of Light from the Divine through the image of the Trinity, through cosmic generation and mundane activity, to Aristotelian corruption.

It is grand because of its dependence on Classical realism and Christian examplarism, its inclusiveness with respect to dialectic as the basis of discovery, hence Guarini's rejection of Parisian physico-mathematics, as mathematics traditionally was never a basis for causality, which it was becoming under Galileo and Descartes. As well as dialectic,

Guarini was also deeply involved in light, as *vincula et via* – a way and a chain, hinting at both Christian and Classical participation – the 'way' of St John's Gospel and Homer's 'golden chain' of light. Regarding this last point, Guarini is unique among writers of *cursi philosophi* of the period 1570-1670 in addressing himself so diligently to the topic of light, and this aspect of his intellectual personality did have important recognitions in his iconography and mathematics, as we will see below. I have demonstrated his use of proportional projection based on the perspectival ladder elsewhere [McQuillan 1991], which supports the design of the dome of the Chapel of the Holy Shroud, Turin (fig. 1).

Fig. 1. The dome of Guarini's Chapel of the Holy Shroud in Turin

As already indicated, it is my claim that the *Architettura civile* faithfully reflects this structure of reality reflected in the sciences, for architecture represented by the geometry of Vitruvius – the Vitruvian trio of projection, the first two, ichnography (the footprint of the building), and orthography (parallel projection) are developed, and scenography is remodelled by Guarini, much as he dismissed it in his *Euclides adauctus,* 1671, a mathematical encyclopaedia. To enter fully into such an interpretation, I will therefore discuss each *trattato* or section of the treatise in turn.[3]

The treatise on architecture

Trattato I established the discipline in its parts, paying some cognisance to Vitruvius in assembling in outline various graphical techniques, which ascend in a chain of intellectual dependency until the apex is reached in Geometry, but with the proviso: *L'Architettura, sebbene dipende dalla matematica, nulla meno e lia un'arte adulatrice, che non vuole punto per la ragione disgustare il senso* [Guarini 1968: Tr. I, Cap. III, 3].

Under *Principi di Geometria necessari all'Architettura*, a condensed introduction to Euclid, he then introduced the propadeutic content, a complete *précis* of both rational and irrational proportion. Thus by bringing proportion to the forefront, Guarini achieved in the *Architettura civile* what he failed to do in the more hidebound treatment of mathematics that is the *Euclides adauctus*. In this spirit he affirmed that *L'architettura pu correggere le regole antiche, e nuove inventare* [Guarini 1968: Tr. I, Cap. III, *Oss. Sesta*, 5].

The second trattato, *Della Ichnografia*, is for Vitruvius the first part of *dispositio*. Already in Trattato I, Cap. II, Guarini warned that: *Il Disegno o Idea, secundo Vitruvio, ha tre parti, delle quali la prima dicesi Icnografia, chela discrizione ed espressione in carta di quello che dee occupare la fabbrica, che si disegna nel piano* [Guarini 1968, Tr. I, 2]. In Euclidean terms this is the line describing the figure; in ontological terms this is the creation of the angels and light on the First Day, and the activity of the soul proceeding from its divine origin; in architectural terms this is the plan and its figuration, which Guarini explores through diverse parts such as stairs and colonnades. This also requires the knowledge of constructing regular figures such as the pentagon and hexagon, in that order. Here the oval and ellipse are also introduced, as well as his argument with the oblique architecture of Caramuel de Lobkowitz, the egregious Cistercian and Spanish count-bishop, where the Theatine discussed circular colonnades.[4]

The third and longest trattato is by far the most instructive as against the more pedantic nature of what comes before and after, as here is found his treatment of the orders. Entitled *Della Orthografia Elevata*, in terms of the Euclidean generation of form, this represents the development of three-dimensional form first from projected elements, and then through their rotation or projection. As Guarini says in its first chapter, *Onde la ortographia certe prime tiene delineazione, diversee forme le sue idee e sono in generale, sorte di sporti detti Projectiones* [Guarini 1968: Tr. III, Cap. I, 73]. Carboneri provides an explanation from antiquity: *Cvr anche Svetonio (De vita Caesarum) che definisce l'ortographia "Formula, ratioque scribendi e Gramaticis instituta* [Guarini 1968: 113]. In ontological terms this represents the production of planar corporiety, and thus Guarini discussed the orders and the articulation of associated arcades and walls – what was also called *ordonnance*. The successor to orthography, central projection (stereography in the *Euclides adauctus*), now appears as linear perspective (*CAPO VIGESIMOPRIMO, Del rendere proportionata la prospettiva che sembra difettosa per cagione della vista*), a discussion of how various optical effects are contrived. This is followed by more oblique architecture, ending with vaults, *la principale parte delle fabbriche*,[5] which in effect was his anticipation of the content of Trattato IV, as the beginning of Trattato III makes clear: *Due sorte di ortografia deve speculare l'archetto; dir per due sono le ortographie, una sielevata, l'altra si chiama depressa; di questa ne scriveremo nel trattato sequente; ora solamente delle prima siano per discorre*. In terms of his cosmological structure, *ortografia elevata* would correspond to the perfect corporiety of the celestial bodies, its influential light and the parallel projection of sunlight, and is therefore related to the perfect ordering of the Orders; the next stage, dealing with vaults, indicated a lower level of embodiment and the introduction of vault construction would thus exploit the full corporeity of crassitude.

In Trattato IV, *Dell'Ortografia Gettata*, Guarini immediately indicated this increased corporeity by relating the utility and necessity of Architecture to the French command of the cutting of stones, *la coupe des pierres* or stereotomy, none of which words he used in the text. Here of course the operations are parallel projection, but the operations are similar to those required for gnomonics, in the scribing on whatever plane sunlight falling on any orientation of dial.

The last trattato was *Geodesia* (Geodesics), and the title of its first chapter openly declared its theme: *Delle transformazione delle superficie piane rettilinee in altre uguali* – metamorphosis and transformation. This theme was underlined by the Bettinian subject of isoperimetry in Cap. VII. Thus Guarini established the first stage of transformation, to be followed by proportional geometrical progression (Cap. VIII) where the progression of forms to infinity can be made equal to a given area, an analogous process to isoperimetry. While such theory, he roundly declared, was not necessary to the architect,[6] it was presumably included to enable the conscientious reader, i.e., one who has also read the *Euclides adauctus*, to follow the St. Vincentian material which constituted *Capo Nono: Della quadrazione, spartimento, ed accrescimento geometrico del circolo* – Gregorius de St. Vincent SJ (1584-1667), who was the well known contemporary circle-squarer who failed in this task.[7] This implied the normal processes of physical development, generation, gnomonical increase or dilation, isoperimetrical equivalence, division and the illusive quadrature, where Guarini threw in a number of simple methods with little elaboration. At last the usual conics of the ellipse, parabola and hyperbola were treated in various forms of transformation and quadrature, all of which are adduced from the *Euclides adauctus*.

The other theme that ran through Trattato V was division, the division of a figure or of land in order to measure the totality of the figure, as geodesics was sometimes defined. This for Guarini meant the final process of mathematical investigation, division to infinity that could only be completed by physical means, i.e., consummation through fire. This may well represent the cycle of elemental transformation, which will be terminated at the end of the world, evoking both the mundane transformation of becoming (the Heraclitan elemental cycle to and from fire) and the teleological destiny determined for the world in Christian eschatology, before the appearance of the new heaven and the new earth of the Apocalypse.

Guarini and Baroque intellectual culture

His Universal Philosophy, and above all his Universal Mathematics, mean much more than the overworked application of 'universal' that was applied to any field of contemporary study embodying conventional encyclopaedic intentions. Not only trained in dialectics and versed in the still, for him, contemplative tradition of mathematics, he was thus heir to these two most distinctive forces in European culture, exactly at the time when these forces were being expressed as never before, in the attempt to create a *mathesis universalis*. While the historical outcome of the period must be judged to have been almost mortally deleterious to the authenticity of these traditions, Guarini was undoubtedly aware of the main lines of the ongoing attack, especially when he rejected the 'physico-mathematics' of the Paris school, of which he had first-hand acquaintance. Nevertheless, within his self-imposed constraints of writing manuals and the bounds imposed by the ecclesiastical and secular powers, Guarini did leave ample evidence of what his world meant, but which, because of the intervening change in our intellectual culture, now requires such effort to recover. Thus his philosophy and mathematics were framed in the old-fashioned didactic manner of his age, time-honoured in the highest way, and his architectural treatise, in contrast, was most unusual but necessarily part of his philosophical world. For Guarini theory and practice were quite separate and only connected, if at all, by the possibility of the arts' communicating essences *demonstrative* or *ostensive*. His interest in astronomy and gnomonics in his examination of the great Book of the World is properly reflected in the Book of the Soul, and can be approached by that paradigm of all Books, the Bible. This is implicit in his *Physica* where proof from scripture is so useful, guaranteeing its universal validity, and its access to truth. As for his mathematics, its ultimate justification must be found in Denys the Areopagite, which accounts for the activities of the 'miraculous Mathematicians' in theological, and therefore the most absolute, terms. These miraculous mathematicians, *contra* Wittkower [Wittkower 1975: 186], are the angelic mediators between the Divine and the human mind, a doctrine well presented in the *Placita*, but ignored or misunderstood by every commentator so far.

Hans Georg Gadamer [1986] has reminded us that mimesis – cosmic mimesis – is still the greatest force in the world of art, and this is what Guarini, in common with all great traditional art, has achieved in his architecture. While a study of his remaining churches lies outside the scope of the present study, it is possible to relate the major parts of the buildings to such an understanding, far more convincing than the weak appeal to 'open-work structures' that was advanced by Pommer [1967], and echoed by almost everyone thereafter. The other noteworthy aspect of their design is the presence of 'proportional projection' in the geometrical consideration of the vaulting and ribbing, though again the object of much witless speculation (Moorish cupolas, etc.). It can be demonstrated that Guarini shared this mathematical understanding with Nicolas-François de Blondel, but in each case, their successors failed to maintain an awareness of this advanced mathematical procedure.[8] Both upheld mathematics as the premier science governing architecture; they shared the Baroque metaphorical structure of space that brings together the different arts and sciences while coming to terms with the practical life, *decorum* and the *ethos* of traditional culture[9] – in other words, 'communicative space'.[10] Thus my assessment is contrary to that of Giulio Argan who insisted in a review of a book on Guarini, that 'no-one more that Guarini has affirmed the non-symbolic, non-allegoric, non-metaphorical character of architectural form', a point of view that is fundamentally antithetical of anything meant by the word 'Baroque' and the work of Guarini.

The grammar of elements

The Theatine, an active philosopher and teacher in the manner of Second Scholastic, was a traditional rationalist, where the certitude of mathematics resided, as Aristotle prescribed, in the sidereal world, an issue that runs through his treatment of mathematics in the *Euclides adauctus*. But he was also influenced by contemporary strains of rationalism, hinted at in his appeal to the authority of Petrus Ramus (Ramée) and Francis Bacon. While not mentioned in any text, another authority was Alberti, and his procedure of identifying elements in his treatise on painting:

> The first thing to know is that a point is a sign which one might say is not divisible into parts. I call a sign anything that exists on a surface so that it is visible to the eye. No one will deny that things which are not visible do not concern the painter, for he strives to represent only the things that are seen. Points jointed together continuously in a row constitute a line. So for us a line will be a sign whose length can be divided into parts, but it will be so slender in width that it cannot be split. A curved line is one which runs from point to point not along a direct path but making a bend. If many lines are joined closely together like threads in cloth, they

will create a surface. A surface is the outer part of a body which is recognised not by depth but by width and length, and also by its properties. Some of these properties are so much part of the surface that they cannot be removed or parted from it without the surface being changed [Alberti 1972: Book I, 37].

While Euclidian in origin, this is evidence that the analogy of syllables and the grammar of orthography – the foundation of the first Liberal Art of the quadrivium, geometry – was a commonplace in European thinking about visual art, which was a subalternate art of the quadrivium. However, it is nonetheless surprising that we find a similar interpretation of building appearing in .the *Architettura civile* in such terms. However there is some further evidence to be considered, that of the treatment of architecture in Claude-François Milliet Dechales SJ's *Cursus seu mundus mathematicum* [1674], a text known to Guarini. The Savoyard Dechales (1621-1678) lived mostly in France and, though not a Cartesian, was abreast of the latest physics, so his inclusiveness points the way to later encyclopaedic efforts to come. This mathematician was followed by C.-A. d'Aviler, pupil of Nicolas-François de Blondel, who modelled his two treatments of the Orders on Scamozzi (1685) and Vignola (1699), indicating a pluralism that became widely rampant.

The importance of Dechales's 'elemental geometrisation' of architecture casts a confirmatory realisation in a work of his great contemporary, Guarini. The instance of trying to invent a basic elemental system of building, through Dechales's rationalisation of the orders, was in accord with the implicit systematisation of the orders proposed by Serlio and confirmed by Vignola, but not adhered to by Guarini, who could invent a variety of orders, as he did in the *Architettura civile*. Guarini went further to organize his treatise on architecture by modelling all five parts on levels of ancient geometrical order, and based where possible on harmonic proportions. As to Claude Perrault, his new system of proportions was not ancient but Cartesian, as placing the architect in charge of nature, and able to dispense with harmonics in favour of authority – the French crown, and custom, which is not clearly defined [McEwen 1998: 325]. Guarini was consistent in applying his learning of Universal Mathematics, mostly to do with harmonics, to his complete presentation of architecture in his overall treatise, and not just the Orders.

Conclusion

It is evident that few great figures of our culture have endured such misrepresentation for so long as has Guarini, some of which I have indicated above. It is to be hoped that this investigation of Guarini's mathematics in the context of the culture of the Baroque will lead to a surer confirmation of his greatness and his intellectual superiority in that 'century of genius', and to place him alongside the great figures of European architecture not because he was like them, but because he understood what he meant to a degree that surpassed them all. Beyond that achievement, Guarini was a thinker who worked hard to explain light, and to support the advance of mathematics through summation to infinity by harmonic proportion, an overall project soon to be solved with the discovery of calculus by the end of the century. With his command of both sacred and secular hermeneutics, Guarini was successful in combining both in repeated ways in this work and architecture.

Finally, no one has attempted to account for the failure of Guarini and his immediate successors to publish the complete treatise on architecture during his lifetime or after his death. The artistic climate in European courts was changing just as Guarini commenced his sojourn in Turin. Through the principal agency of Poussin and the Fréart brothers across Europe, there was a campaign to reject the *licence* of such figures as Borromini and Bernini, measured in the alleged 'conversion' of the aged sculptor eventually to their form of artistic rectitude in Rome and also in Paris. This acerbic classicism which successfully alleged that Borromini was a Goth, may in due course have won out at the court of Turin, due to the supreme variety of Guarini's Orders, including the Salomonic and even 'undulating architecture' not to mention a column version of the hated Gothic, all of which figure so prominently in the plates to the *Architettura civile*. Could these have now caused offence, after the completion of the Royal Chapel of the Holy Shroud? Could these plentiful licenses in Guarini's treatise have been too much for any prospective high-ranking patron to support in a changing intellectual climate, caught between Rome and Paris? Thus it was left to another generation, when the promising and well-connected young architect Bernardo Antonio Vittone was successful in bringing the full work on architecture to the light of day in 1737, with the prominent support of the Provost General of the Theatines in Rome who supplied the preface. In Vittone's hands the magic of the Baroque had its Indian summer in Piedmontese towns where he built so many parish churches, but with his death, the vigour and vision of the Baroque was seen no more south or even north of the Alps. Neoclassicism had won, with the secularisation of all artistic forms to follow.

Notes

1. *Conclusio: nam omnes artes vel a Mathematica, vel a Philosophis, vel a medicina dependent; quae omnes scientiae vel rerum similitudinem, vel proportionem, vel convenientiam considerant* [Guarini 1665, 214].
2. Pérez-Gómez has pointed out that the treatise is modelled on that of Carlo Cesare Osio, a Milanese architect, with the same title, published in 1684, and he indicates that this was a model for Guarini's deliberations. However by the time of Osio's publication Guarini had died, and I suspect that they knew each other, Guarini actually dying in Milan, where Osio might have been his assistant in his last days. Therefore the proportional doctrine of Osio may be from Guarini, so he did not need to quote ancient sources, as Pérez-Gómez suggests he should, and which was uncommon in many mathematical texts anyway.
3. In his paper 'Guarini the Man' [1975], Rudolf Wittkower did not examine the deeper mathematical meaning of the *Architettura civile*.
4. [Guarini 1968: Tr. II, Cap VIII, *Del Modo di disporre un colonnato nel tondo*]. Guarini opposed the Benedictine's consistent corrections of the apparent sizes of elements viewed from a distance, which indicates that the spectator is not immobilised, for Guarini.
5. [Guarini 1968: 277]. Guarini was certainly very conscious of Gothic achievements, and the Chapel of the Holy Shroud, with its iron reinforcement as evidence of techniques arising in Paris at that time.
6. Guarini was surely describing the culture of Italian and French architects in his day, which no one has yet indicated, cf. [Guarini 1968: Tr. I, Cap. III]: *quando però si tratta, che le sue dimostrazioni osservare siano per offendere la vista, la cagione, la lascia, ed infine contradice alle medesime* (...so that, if the eye should be offended by the adherence to mathematical rules – change them, abandon them, and even contradict them). Presumably the background notion is to invent new rules.
7. As stated by one of his students, Saint Vincent was 'one of the great pioneers in infinitesimal analysis'. J. E. Hofmann, 'Saint Vincent, Gregorius', in voce, [Gillispie 1970-1981]. His main work was *Opus geometricum quadraturae circuli et sectionum coni*, 1647.
8. I will explore this connection in my forthcoming paper, 'The Mathematics of Nicolas-François Blondel and the Four Problems of Architecture'.
9. 'The unity of Baroque space . . . is established by the metaphorical structure of space, which has the capacity to hold together different arts and at the same time meet all the most important conditions of practical life: *decorum* and *ethos*' [Vesely 1987: 33].

10. [Vesely 1994: 81]. The author explains the concept of communicative space with reference to the Baroque library on pp. 80-81.

Bibliography

ALBERTI, Leon Battista. 1972. *On painting; and On sculpture: the Latin texts of* De pictura *and* De statua, edited, with translations, introductions and notes by Cecil Grayson. London, Phaidon.

DECHALES, Claude-Francois Milliet. 1674. *Cursus seu mundus mathematicus.* Lyon: Anisson, Posuel et Rigaud.

FAGIOLO, Marcello. 1970. La *Geosofia* del Guarini. Pp. 179-204 in vol. II of *Guarino Guarini e l'internazionalità del Barocco. Atti del convegno internazionale, 30 September-5 October 1968,* Torino: Accademia delle Scienze.

GADAMER, Hans Georg. 1986. *The Revelance of the Beautiful and Other Essays.* Translated by Nicholas Walker and edited with an Introduction by Robert Bernasconi. Cambridge, Cambridge University Press.

GUARINI, Guarino. 1665. *Placita philosophica.* Paris, apud Dionysium Thierry.

———. 1671. *Euclides adauctus et methodicus mathematicaque universalis.* Turin, Zapata.

———. 1964. *Architettura civile,* (Torino, 1737). 2 Vols., edited by Bernardo Antonio Vittone. London, republished by the Gregg Press.

———. 1968. *Architettura civile* (Torino, 1737). Nino Carboneri, ed., Milan: Il Polifilo.

MAZZEO, Joseph Anthony. 1978. *Varieties of Interpretations.* Notre Dame: University of Notre Dame Press.

McEWEN, Indra Kagis, 1998. On Claude Perrault: Modernising Vitruvius. Pp. 321-337 in *Paper Palaces, the Rise of the Architectural Treatise.* Vaughan Hart and Richard Hicks, eds. Yale University Press, New Haven and London.

McQUILLAN, James. 1990. The Mathematical Places of Mario Bettini. *Scroope the Cambridge architectural magazine,* Issue no. 2, (June, 1990): 18-23.

———. 1991. Geometry and Light in the Architecture of Guarino Guarini. Ph.D. Thesis, University of Cambridge. Copy in the British Library.

MEEK H. A. 1988. *Guarino Guarini and his Architecture.* New Haven: Yale University Press.

PÉREZ-GÓMEZ, Alberto. 1983. *Architecture and the Crisis in Modern Science.* Cambridge, MA.

POMMER , Richard. 1967. *Eighteenth-century Architecture in Piedmont: the open structures of Juvarra, Alfieri & Vittone.* New York, New York University Press.

VESELY, Dalibor. 1987. The Poetics of Representation. *Daidalos* **25** (Sept. 1987): 24-36.

VESELY, Dalibor. 1994. Communication and the Real World. *Scroope* **6** : 78-81.

VESELY, Dalibor. 2004. *Architecture in the Age of Divided Representation: the Question of Creativity in the Shadow of Production.* The MIT Press, MA.

WITTKOWER, Rudolf. 1975. Guarini the Man. Pp. 177-186 in *Studies in the Italian Baroque.* London, Thames & Hudson.

About the author

James McQuillan was brought up in Ireland where he was trained as an architect at University College Dublin, graduating in 1969. He proceeded to study the M.A. in the History and Theory at the University of Essex where he was taught by Profs. Joseph Rykwert, Dalibor Vesely and George Baird, where Daniel Libeskind was one of his fellow-students. Returning to Ireland, he was engaged in professional practice and part-time teaching in his old school of architecture, as well as spending a year in Rome as a *borsista* of the Italian Government. Elected a member of the RIBA, from 1978 to 1982 McQuillan ran his own practice in Northern Ireland, and then he went to teach in Saudi Arabia, in Dammam. In 1986 he arrived at the University of Cambridge where Dalibor Vesely supervised him in his thesis on Guarino Guarini, and he has since been teaching and researching in architecture worldwide, with contributions to the international conference circuit. His book, *Modern Architecture as Landscape,* is being completed and he teaches part-time at Lincoln and Nottingham Trent Universities.

Paolo Napoli

Dipartimento di Ingegneria
Strutturale e Geotecnica
Politecnico di Torino
Castello del Valentino
Viale Mattioli, 39
10125 Torino ITALY
paolo.napoli@polito.it

Keywords: Guarino
Guarini, Chapel of the Holy
Shroud, structural
mechanics, masonry
construction

Research

A Structural Description of the Chapel of the Holy Shroud in Torino

Abstract. The structural investigations performed on Guarini's Chapel of the Holy Shroud in Torino have made it possible to come closer than ever before to understanding its structural behaviour. They have also shed light on the building's history. This paper presents some of the new findings about key elements of the structure, and mentions some of the still open questions.

Guarino Guarini was appointed Ducal Engineer for the Chapel of the Holy Shroud on 19 May 1668. At that time work on the construction of the temple destined to host the Holy Shroud had been ongoing for some 60 years. The earliest document that refers to the shipment of black marble columns for the chapel is dated 1607; this refers to the construction of the first design, prepared during the first years of the 1600s by Carlo di Castellamonte. It can be seen that the floor of the new temple was located on practically the same level as the Cathedral, separated by just a few steps (fig. 1).

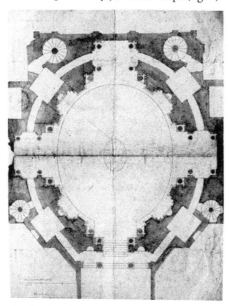

Fig. 1. Castellamonte's plan for the Chapel of the Holy Shroud (1621, Fondazione Umberto II e Maria Josè di Savoia)

Documents about the progress of the construction testify to work on the site up to 1621; then follows a period of inactivity until, at the request of Prince-Cardinal Maurizio a new design was prepared by Amedeo di Castellamonte, son of Carlo, and by the Italian-Swiss architect Bernardino Quadri. Quadri's design was adopted; an official warrant

Nexus Network Journal 11 (2009) 351–368   NEXUS NETWORK JOURNAL – VOL.11, No. 3, 2009 **351**
DOI 10.1007/s00004-009-0003-y; *published online* 5 November 2009

authorising its construction exists. A wooden model of the project was prepared, but neither the model nor Quadri's drawings have survived. There is, however, a plan in Turin's Biblioteca Reale which corresponds to descriptions of Quadri's design.

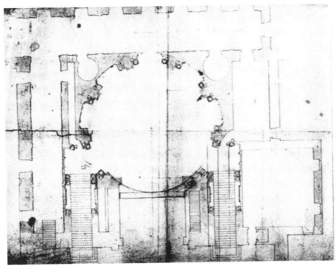

Fig. 2. Plan of Quadri's Chapel of the Holy Shroud (ca. 1655, Biblioteca Reale di Torino)

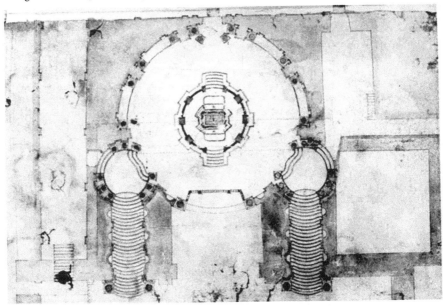

Fig. 3. Plan of Guarini's Chapel of the Holy Shroud (after 1694, Archivio Capitolare , Torino)

Quadri's new project is radically different from the earlier one by Carlo di Castellamonte (fig. 2). The shape of the chapel is now circular and not oval, and the floor of the new chapel is raised to the level of the *piano nobile* of the Royal Palace, that is, some 5.50 metres above the level of the Cathedral. Access to the Chapel was by means of two long lateral stairways.

Builder Bartolomeo Pagliari was appointed in 1657 to demolish – as much as necessary – the 1611 foundations and to proceed with the new construction. In 1666 – two years before Guarini's arrival on the scene – the walls had reached the height of the second order. However, there were doubts about whether or not the quality of the materials and the construction completed thus far were adequate to support the structure of the dome that was planned. A special meeting took place at the site on 10 September 1665, and included engineers, architects, builders and surveyors.

When Guarini assumed his position in 1668, he took over work on a project that was considerably advanced, with a number of features that could not be dispensed with: the circular plan, the lateral stairways, and the large window that opened into the Cathedral. He dismantled part of the uppermost structure – down to the level of the first order – and radically redesigned the rest (fig. 3).

From the point of view of structure, it is important to note that Guarini's building is supported by Quadri's foundations. Guarini limited himself to strengthened some crucial aspects of the structure, filling in the voids of the two circular stairs in the east wall and of the stair that was located in the middle of one of the walls of the Cathedral.

A comparison of a survey of Guarini's Chapel with the designs of Quadri and Carlo di Castellamonte makes it possible to see how the designs of Guarini and Quadri are located in the same position, while the design of Castellamonte regarded a much larger area, projecting well into the interior of the courtyard of the royal palace (fig. 4).

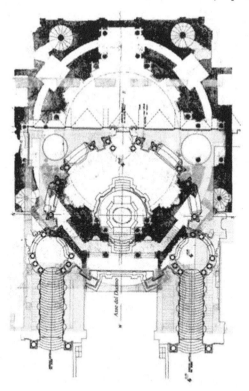

Fig. 4. Overlaying Guarini's plan of the Chapel over the design of Quadri makes the similarities and differences evident

During the course of investigations of the project for the restoration of the structure of the Chapel following the 1997 fire, a number of borings have been made beneath the surface of the courtyard of the royal palace. In effect, in positions that correspond to the locations of the masonry walls of the Castellamonte design have been found foundation structures that extend down to a notable depth. Chemical examinations on the borings extracted have confirmed that they date from several centuries ago.

In fact, the concerns about the capacity of the basement walls to support the weight of a high dome which gave rise to the special site meeting of September 1665 were well founded. Guarini himself had the most significant voids filled – those of the east wall and of the Cathedral – but the fills were by necessity executed using refuse materials that were only slighted mixed with cement. The borings executed have clearly showed the inconsistency of the masonry (fig. 5).

Fig. 5a. Core extracted from the wall of the Cathedral

Fig. 5b. Core obtained by the wall of the Chapel in the zone of the right circular staircase

In spite of this infilling the stress in the external part of the walls remain very high, in excess of than 1 MPa, compared to a resistance of the wall of some 1.2 – 1.4 MPa. This fact emerges from both the modelling of the finite elements and the measurements taken on site with flat jacks.

The walls of the upper part of the Chapel, entirely due to Guarini, do not present problems of resistance. The borings extracted have revealed a uniform composition of masonry, even though not of excellent quality, with an external veneer in Frabosa stone of a thickness on the order of 15-20 cm.

As is often the case with Guarini, there arise many uncertainties and problems in identifying the effective structural system, distinguishing it from elements that appear structural but are essentially decorative.

A first example of this difficulty is found in the second order. Here are immediately evident the three marble arches that project beyond the marble surface and define the three pendentives that rise to the level of the ring of the balcony of the drum. The formal and symbolic significance of the pendentives, related to the partitioning of the plan of the Chapel into nine parts, of the drum into six parts and of the cupola into twelve parts, has been noted for some time (fig. 6).

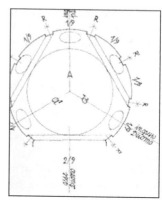

Fig. 6. View of the interior of the Chapel of the Shroud (left), and diagram of the division plan into nine parts (Passanti, 1963)

To the formal significance there is no corresponding true structural significance. A first examination of the original drawing by Guarini, held in the Biblioteca Apostolica Vaticana (1686) shows that the pendentives are placed in areas that are essentially empty, and that they are incapable of resisting the notable weight that they ostensibly transmit.

The measurements taken with the flat jacks have confirmed the marginal role played by the three marble arches.

However, there are four masonry arches, arranged about an irregular rectangle and placed in areas of thick masonry construction, that play a significant role (fig. 7). The two orthogonal ones are relatively visible on the axis of the Cathedral, towards the Royal Palace and towards the Cathedral itself.

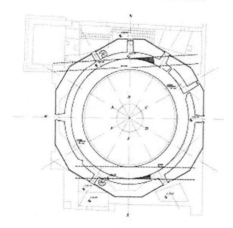

Fig. 7. Plan section (left) and photograph (right) of the masonry arches

The other two, roughly parallel to the axis of the Cathedral, follow the direction shown in the plans. These arches are visible in the passageway beneath the loggia, beyond which small segments of them project.

The precise structural role played by these last two arches is open to question. They are in fact completely incorporated into the masonry, without there being an opening below them. As a consequence they should behave as arches – receiving the loads and transmitting them to the imposts – only in the event that the masonry underneath them was to settle. When there is no such settlement (and there are no visible indications that such settlement has occurred), the vertical compressions in the body of the walls would substantially pass unperturbed through the thickness of the arch and be transmitted to the masonry below. More or less the same can be said of the possible horizontal forces, especially those of compression as in the upper part of a dome. As a consequence, when modelling the structure, the presence of these arches incorporated in the masonry can be neglected.

It must be borne in mind that the insertion of arches within the fabric of the masonry, without an apparent necessity in terms of statics, occurs frequently in the masonry of the chapel, so much so that it appears almost as the application of a structural rule of thumb. Identification of these arches where the masonry is covered by plaster is practically impossible.

On the other hand, the system of metal ties that are present at all levels of the Chapel is very important. In some cases these correspond to precise requirements of resisting horizontal thrusts, such as in the case of the ties at the bases of the arches at the level of the oculi at the level of 20 m.

In other cases these play the more generic role of tying together the wall fabric, without there being evidence of local thrusts.

In still other cases these ties are required to sustain elements that lean or project, such as the cornices of the second level (the *bacino tronco*) and the three marble arches, a further confirmation of the marginal structural role of the arches.

The ties and the metal rings naturally play an increasingly important role as the thick masonry structure of the lower level rises towards the slender structure of the loggia and cupola.

A fundamental structural role is played by the ring placed in correspondence to the cornice of the drum. This is difficult to see because it is camouflaged by the mullions of the great windows.

The essential role of the rings is connected to the unique typology of the Chapel of the Shroud. In traditional typologies, at the base of the cupola there is always a continuous masonry ring, which resists the radial thrusts of the cupola (fig. 8). Often this masonry contains a metal tie ring. In any case, even where the metal tie is not present, the horizontal traction resistance of the masonry (which is not zero, due to the offsets of the courses and to friction generated by the vertical compressions) is an effective resistive force.

Fig. 8. The masonry ring that usually resists the outward thrust of a dome. Comparison of the dome of St. Peter's (left) with that of the Chapel of the Holy Shroud (right), where the masonry ring is interrupted by the arches of the large windows (indicated by the arrow)

In the Chapel of the Shroud, the masonry ring at the base of the cupola is broken by the six great windows of the drum, and cannot therefore exert any forces of containment. This role is entirely entrusted to the metal rings.

During the fire of 11 April 1997, burning hot smoke issued from the great windows. The rings of the drum were broken in two places, and as a consequence a series of serious lesions occurred in the arches, the ribs, and at the stone vaults of the drum (fig. 9).

Fig. 9. left, the hot smoke coming out of the windows during the fire; middle, the broken ties

Settlement was due to the outward displacement of the pilasters of the drum, which form a frame with the architraves and the stone vaults. The rapid placement of temporary stirrups and of a pre-stressed ring made of four cables prevented further settlement (fig. 10).

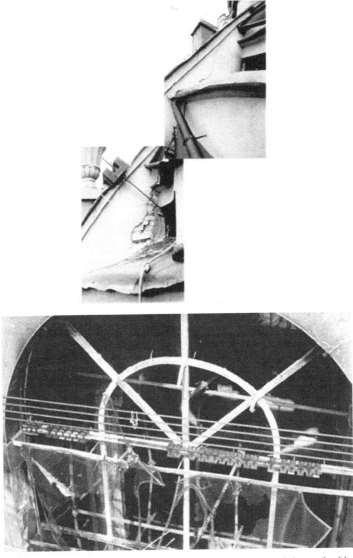

Fig. 10. Temporary stirrups placed after the fire, on the ribs (above), and the steel cables within the windows (below)

Let us now turn to the most complex and mysterious part of Guarini's Chapel, that is, the open-work dome known as the "basket" (*cestello* in Italian) (fig. 11). From the interior this appears to be formed of six series of superimposed marble arches that are separated from one another by cornices, also of marble. Each series is formed of six arches arranged in plan in the form of a hexagon, rotated with respect to those above it so that the impost of the upper arch falls on the keystone of the one below. In essence, the vertices of the hexagon correspond to the halfway point of the sides of that below it, and hence the offset between one series and the next corresponds to a rotation of 30°. Because of this, going up from one level to the next the radius of the circumference circumscribed and the opening of the arches is reduced by a coefficient of cos30°, that is, 0.866.

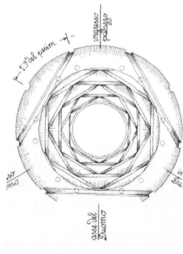

Fig. 11. Photograph and drawing of the "basket" dome of the Chapel (Passanti, 1963)

In actual fact, on the exterior of every stone arch there is a masonry arch that exactly follows its profile; the stone arch and the masonry arch are connected to each other by numerous metal elements, sealed to the stone with molten lead; in the cornices that separate the series of arches there are ties that follow the whole hexagon.

Seen from the exterior, the series of superimposed arches are less recognisable, in part because they only partly follow the shape of the window, while on the other hand the elements most in evidence are the ribs that lie on the meridians of the cupola and converge towards the lantern, alternately placed at the location of the pilasters and on the keystones of the arches of the great windows (fig. 12).

Fig. 12. Interior (left) and exterior (right) of the "basket" dome of the Chapel (Passanti, 1963)

To this double image – interior and exterior – correspond two contradictory interpretations of the structural behaviour of the cupola.

The first of these sees the load-bearing structure in the union of the superimposed arches and the ties present in the cornices, attributing a minor role to the external ribs, which support the urns: to this interpretation I give the name "the arch system" (fig. 13a). The second interpretation, of long tradition (M. Fagiolo 1968) but recently brought back into discussion, recognises the load-bearing system exclusively in the twelve external ribs and in the corresponding points of vertical continuity on the interior, demoting to pure appearance the role of the superimposed stone arches and the corresponding masonry arches: to this interpretation I give the name "the rib system" (fig. 13b).

Fig. 13. The two systems used to interpret the load-bearing structure of the basket dome. a (left), the arch system; b (right), the rib system

Since the two structural schemes are both effectively present, it is evident that both theories can be have a basis in truth. In order to identify the most reliable hypothesis, it is therefore necessary to examine the problem from multiple points of view:

- the structural analysis of the monument, by means of modelling of the finite elements;
- investigation through measurements of the actual distribution of forces;
- archival research;
- detailed examination of the construction techniques used in the various parts of the cupola, deducing the structural significance from the technique used;
- examination of the distribution of the cracks caused as a consequence by the fire of 1997.

The refinement of the instruments currently used for structural analysis makes it possible to formulate highly reliable models of the structure of the Chapel (fig. 14a), attributing to the elements constitutive relationships that are both linear and non-linear. In both cases, the results of the analysis confirm what structural intuition suggests: the rib system, employed substantially in axial resistance, is more rigid than the system of superimposed arches and their related ties in eliminating the thrusts. As a consequence the thrusts are channelled essentially through the twelve ribs, leaving only a modest role to the arches.

However, the analysis of finite elements furnishes important information regarding the simple "rib system". In fact, it turns out that the ribs do not all contribute equally to the load-bearing; rather, the so-called "long ribs", those placed in correspondence to the location of the pilasters of the drum take on the major part of the load, leaving only a modest portion to the "short ribs", that is, those placed on the keystones of the arches of the great windows. This too could be intuited, by comparing the rigidity of the support of the pilasters with the more yielding arches of the great windows.

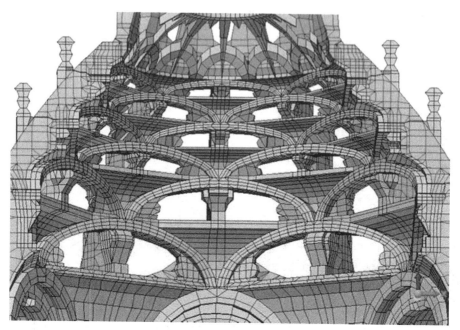

Fig. 14a. Part of the structural analysis performed on the chapel

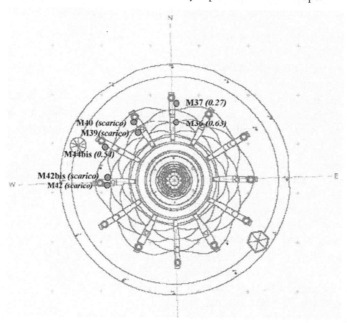

Fig. 14b. Compression stress in the ribs measured using flat jacks

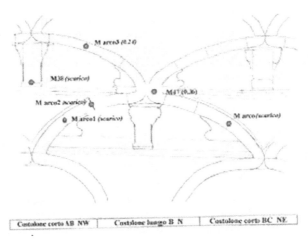

| Costolone corto AB NW | Costolone lungo B N | Costolone corto BC NE |

Fig. 14c. Compression stress in the marble arches measured using flat jacks

As part of the structural investigation relative to the rehabilitation project, measurements have been taken of the compressive forces present in the ribs as well as in the arches. The results show a mean compression on the order of 0.6 MPa in the long ribs, and almost zero in the short ones. The marble arches and the masonry arches parallel to these turn out to not be carrying any load (however, the measurement on the marble arches is not very reliable, as will be explained below).

Fig. 15. Sections of the Chapel by Guarini (left) and that by Borgonio (right)

In spite of these results it would be erroneous to reach the hasty conclusion that the external ribs represent the primary structural system of the Chapel, and that the system of superimposed arches play a role that is negligible in terms of statics. Above all it would be erroneous to believe that Guarini conceived a complex spatial system of marble arches of notable thickness without entrusting to these a primary structural role. There is much evidence for these assertions, first among them the iconographic documents.

There exist two drawings that are coeval with the Chapel of the Shroud, one due to Guarini himself and the other attributed to G.T. Borgonio (1669-1670) (fig. 15). In neither of these drawings do the ribs show the vertical continuity that is indispensable in order for them to function as arches. In the drawing by Guarini, in place of the ribs there are large steps that support the urns.

In a recent publication [Dardanello 2006] the importance of the only known drawing by Guarini for the Chapel of the Holy Shroud has been shown (fig. 16). Making reference to the area corresponding to a side of the hexagon, Guarini clearly defines both the geometric scheme and its function in terms of statics: neither modillions nor ribs are shown, and the only load-bearing structure is the system of superimposed ribs.

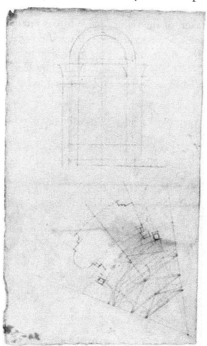

Fig.16. Study of a segment of the drum and dome for the Chapel of the Holy Shroud (G. Guarini, ca. 1675, Archivio di Stato, Torino)

Other evidence confirms that the original concept for the dome did not call for ribs. After the 1997 fire, the plaster of the ribs, which even before the fire had flaked off in large areas, was completely removed, allowing the fabric of the masonry of the ribs to be surveyed and photographed (fig. 17). It is thus possible to note several very significant aspects:

- The ribs were originally built with large steps as shown in the drawing by Guarini. The curved profile, added later, besides being only roughly adapted to the pre-existing profile, is not embedded or tied into this;
- The fabric of the masonry, in addition to being of the poorest quality, is laid in horizontal courses. It is thus clear that these elements were not constructed to function as arches, and do not therefore play the role that the "rib system" attributes to them.

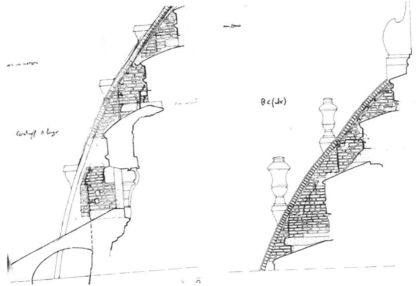

Fig. 17. The masonry of the ribs showing the stepped profile originally planned by Guarini and the curved profile added later

The imposts of both the long and the short ribs occur on the radial brick arches, which are slender and realised with a masonry of mediocre quality.

Embedded into the ribs, between the courses, are long stone elements (the so-called *mensole*, or brackets) that emerge in the interior of the Chapel to form modillions, and appear immediately beneath the imposts of the marble arches.

In stark contrast to the very poor quality of the construction of the ribs is the care with which the arches of the cupola have been realised, both those in masonry and those in marble. The brick courses and the stone joints are correctly oriented in order to behave as arches; the masonry arches and marble arches are joined to each other by a number of connections in iron (up to six for each stone) sealed into the marble with lead and incorporated into the masonry; the joints between individual pieces of marble are strengthened by iron wedges. Either the two sets of arches were constructed at the same time, or those in marble were constructed before the masonry ones on the exterior (the configuration of the metal connectors provides certainty of this).

The rib's behavior after the fire also confirms that they were not conceived as a load-bearing system. Immediately after the fire it was possible to see that large cracks had formed in the long ribs (see fig. 10), followed by minor lesions in the short ones. All of the stone brackets that join the long ribs to the modillions are broken, while none of those of the short ribs are. This behavior can be explained if we assume that, before the fire, the weight of the dome was essentially borne by the system of arches, and only in small part by the ribs. Because of the effects of the high temperatures due to the fire, the hexagonal metal chains present at each level of arches probably lost their rigidity; because of the sagging that followed in consequence, the stone arches transferred a great part of the weight they carried to the ribs, via the modillions and brackets. It was mainly the long ribs, more rigid, which were the first to crack and whose modillions are all broken, and to a lesser degree that of the short ribs.

In spite of this, the system of arches still performs a limited load-bearing function, as has been ascertained in the course of a test in which a half-arch was disassembled. During the test, as soon as the marble arch was removed, the instruments placed on the masonry arch behind it (previously taken out of compression and with open joints) registered the onset of a compression force, a consequence of the loads borne by the marble arch being transferred to it. During disassembly, it was ascertained that the contact between the marble blocks, because of the presence of iron plates, was present exclusively in a small area very close to the masonry: for this reason in tests with flat jacks, necessarily performed on the outermost joints, an absence of compression was registered.

For all of these reasons, it must be concluded that the dome built by Guarini had to have been based on a system of arches, even though now, after the fire, the dome is primarily supported by a system of ribs.

But if Guarini's intention was to construct the "arch system", then what led to the curved profile of the ribs that was constructed successively? Geometry, a science in which Guarini was a master, can contribute to the solution of this enigma.

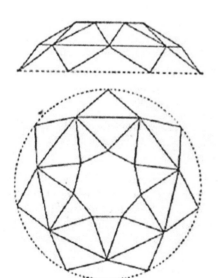
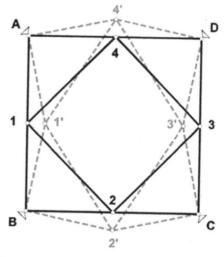

Fig. 18. a (left), a diagram of the cupola of the chapel abstracted to form essentially a grid; b (right), possible movements of the nodes of the grid

If we imagine the cupola of the Chapel as being formed by superimposed arches and hexagonal cornices, but without ribs, and we consider the connections at the foot and at the keystones of the arches as hinges, we obtain a typology that is substantially identical to a grid cupola on a hexagonal base (fig. 18a). These cupolas have a unique property:

they are weak if the number of sides of the base (n) is even. (Even when n is an odd number and not equal to 3 the structure is still badly connected.)

We can understand the nature of the weakness if, for simplicity's sake, we consider only one order of rods on a square base. There remains the possibility of movement of nodes 1 and 3 towards the interior and nodes 2 and 4 towards the exterior (fig. 18b).

The role of the ribs might therefore be that of assuring the stability of the whole rather than that of absorbing the vertical loads. This would explain the horizontal coursing of the masonry fabric, unsuitable for functioning as an arch but perfect for giving stability to the arches that were intended to resist the vertical loads. While not identical, the role of the ribs is analogous to that played by the *frenelli* (wing walls used to help the vault maintain its shape) in barrel vaults, which are in fact laid in horizontal courses.

According to this hypothesis, the dome did not initially feature modillions at the keys of the arches, nor were the external ribs shaped with a continuous curvature, but rather with simple steps as shown in the first designs (see fig. 17). The modillions and the continuously curved profile of the ribs would have been added at a later time, in order to improve overall stability

A confirmation of this hypothesis can be found in the constitution of the cupola. Looking at the intrados of the marble arches, it can be seen that the surface is worked in bas-relief, as is the part covered by the modillions (fig. 19). This means that those parts were intended to be seen, because, naturally, surfaces intended to remain out of view are not normally decorated.

Fig. 19. The partial destruction of a modillion reveals decoration in bas-relief on the intrados of a marble arch even on the part covered by the modillion

Thus it is reliably certain that the dome of the Chapel of the Holy Shroud was conceived by Guarini and initially built without the modillions at the keys of the arches and with simple steps on the exterior in place of the ribs we see today. In a later moment, either in the course of construction or after its completion, indications of weaknesses led steps being taken to improve of the overall structure, inserting the modillions and transforming the steps into continuously curved ribs.

One possible hypothesis is that this occurred at the time when, in a letter dated 8 May 1681 from Marie Jeanne of Savoy-Nemours, then Regent of the Duchy of Savoy, to the Duke of Modena, Guarini was urgently summoned as the only one capable of bringing the work to completion.

A careful examination of how the dome is constituted provides us with a further element of evaluation.

Examining the interior of the succession "impost of the upper arch to keystone of the lower arch to modillion, etc.", it can be seen that, according to the "rib system", at the precise point where the load of the arches should be transferred to the ribs through the modillions, that is, at the point of contact between the arch keystone and the modillion, the marble elements do not touch each other (fig. 20). Instead, there is always a layer of mediocre-quality masonry measuring between 8 and 10 cm.

Fig. 20. Detail showing "point of contact" between the keystones of the arches and the modillions which are actually separated by a layer of poor quality masonry

What is the significance of this layer of masonry? It is hard to believe that it is simply to absorb construction tolerance, when the tolerance of all the other marble elements is on the order of a centimetre.

It is possible to formulate two hypotheses (which are not necessarily mutually exclusive). It is first of all evident that, because the modillions had to be inserted within a marble structure that was already built, a sufficient amount of "play" in order to mount was necessary: for this reason the modillions are lower than they should be in theory, and later the gap was simply filled with masonry. But it is also possible that there is a more refined reason. In this hypothesis Guarini, finding himself forced to insert the modillions for reasons of stability, but wanting to avoid the possibility of their transferring any vertical loads to the ribs, which had not been conceived as arches, intentionally left a void between the modillions and the arches, so that the primary load-bearing system assumed the deformations of the vertical loads, including those consequent to short-term settling, and only later filled the gaps with masonry in order to assure overall stability.

This hypothesis gives due credit to Guarini, who most certainly knew how to construct arches as arches and who would never have constructed ribs destined to carry vertical loads with horizontal coursing. In all likelihood he also knew how to introduce disjunctions or interstices into the structure in order to avoid improperly burdening with vertical loads elements that were purely intended as stabilisers. In actual fact, due to the vicissitudes of the building and in particular to the fire, this is what has happened, resulting in the inevitable cracks and breakages.

In conclusion, the complexity of the structural concept of the Chapel of the Shroud and of the story of its construction do not permit approximate simplifications and superficial schematisations. However, accurate observations of the constructive techniques can allow us to come ever closer to understanding Guarini's structural concept. At the same time, modern investigative means, both on-site and numeric, furnish precious information regarding the actual behaviour of the building. Both kinds of information must be thoroughly acquired before beginning any intervention whatsoever of rehabilitation, which must tend towards a restoration of the original structural concept and which must certainly not be limited to satisfying requirements of statics by means of the artificial reinforcement of elements not originally conceived to serve load-bearing functions.

About the author

Paolo Napoli, born in 1949, is full professor of Structural Design at Politecnico di Torino, Faculty of Architecture. A relevant part of his research activity is devoted to the evaluation of safety and the rehabilitation of existing structures, both in reinforced concrete and in masonry. As a structural consultant, he has been involved in the rehabilitation of important historical buildings. Presently he is appointed to the design for the structural rehabilitation of the Guarini's Chapel of the Holy Shroud.

Ugo Quarello

Via Figlie dei militari 1 bis
10131 Torino, ITALY
ugoquarello@gmail.com

Keywords: Guarino Guarini,
masonry structures, Baroque
architecture, Chiesa di San
Lorenzo

Research

The Unpublished Working Drawings for the Nineteenth-Century Restoration of the Double Structure of the Real Chiesa di San Lorenzo in Torino

Abstract. In his church of San Lorenzo in Torino Guarino Guarini conceived a structural organism that transfers the weight of the dome to the foundations by means of a partially concealed ribbed structure. This rare example of architecture in which form and structure do not coincide is not the result of an eccentric choice of technical virtuosity, but rather a structure mechanism aimed at achieving certain effects that have been knowingly sought out by the architect.

Introduction

The church of San Lorenzo is the product of the profound cultural and spiritual formation as well as the genius of the Theatine priest Guarino Guarini: it is an example of an architecture that is one of a kind. Its proportions, the constructive system, the use of light as a constructive material and the particular dome with its interwoven arches, all contribute to making Guarini's building one of the greatest masterpieces in the history of architecture. What makes the constructive system that Guarini used special is his conception of a structural organism that transfers the weight of the dome to the foundations by means of a partially concealed ribbed structure. The visitor entering the church does not perceive the actual structure but only the shell that masks it: the dome appears to be resting on the drum, which rests in its turn on four spherical pendentives supported by four arches that would seem to transfer the entire weight of the structure to the ground by means of eight slender columns in red marble. In reality, a complex framework of arches and vaults, hidden from view by the shell itself, performs the load-bearing role. What we have here is a rare example of architecture in which form and structure do not coincide. This is not the result of an eccentric choice of technical virtuosity, but rather a structure mechanism aimed at achieving certain effects that have been knowingly sought out by the architect. Guarini, a man of great scientific and theological knowledge, a mystic and experimenter, mathematician and unflagging researcher, conceived of architecture as an opportunity to manifest the religious tending towards the Divine and believed that constructive techniques were legitimate means for astonishing and arousing a sense of wonder.

1 The morphology of the double structure of the Real Chiesa di San Lorenzo

As the tradition of vaulted structures of domes in churches developed, one particular solution made it possible to place a smaller dome over a larger lower area: this involves the use of pendentives, segments of spheres, which transfer the weight of the dome directly to the vertical walls at the base. This system is significant formally as well as structurally.

Guarini had particular objectives in mind as he composed the interior space of the church, such as effects of light and shadow, emptying of volumes, concatenation of elements in order to create a sense of dynamism as well as astonishment or wonder, etc. For these reasons, having studied Arab and Byzantine structures, he created a structure that was quite distant from traditional schemes and which, given its dimensions, surpassed even the most complicated creations of Moorish Spain.

The classic structure with its frescoed pendentives, drum and dome appear in the church of San Lorenzo, but, as was alluded to earlier, this is more apparent than substantial. The church shows innumerable inconsistencies between the elements that make up the interior and the static reality of the structure. We can find details that leave us perplexed with respect to a traditional composition. The dome appears rest on the drum (though this is perforated by eight windows); the drum appears to rest on the four spherical pendentives (which appear solid but are actually, unbelievably, empty); the pendentives appear to be supported by four arches, although these are pierced through at the impost and the keystone where they would require the most strength; the arches appear to transfer the entire weight of the structure to the ground by means of eight very slender columns in red marble. These details, when considered carefully, belie the fact that what we see is actually an interior shell that goes from the pavement up to the impost of the dome, and which covers the actual load-bearing structure.

On the other hand, from the level of the impost of the dome it is possible to see the load-bearing structure, but the double structure is still present, and in this case is manifested by the external band with windows that surrounds the dome.

It should be noted that the secondary, non-load-bearing structures do have, in any case, a structural function, in that they help stiffen the principal scheme. Thus for example the drum is the element that collects and redistributes the loads between the vaulted system of the dome and the framework of the principal arches below.

Summarising, we can list the elements of the main load-bearing structure of the church, starting with the highest (fig. 1):

A. The small upper dome with its drum (this does not appear in the figure);
B. The ribs of the dome;
C. The four main arches;
D. The four diagonal arches;
E. The small corner towers;
F. The double diagonal climbing arches;
G. The structure of the four Serlian windows;
H. The four conical vaults;
I. The perimeter walls at the base stiffened with pilasters.

In an analogous way, we can list the elements of the interior shell:

1. The vaulted shells between the ribs of the dome (these do not appear in the figure);
2. The masonry external crown with corner nodes exterior to the dome (this does not appear in the figure);
3. The drum with the ellipsoid windows beneath the dome;
4. The four conical arches of the Serlian windows;
5. The four spherical pendentives;
6. The band above the arches of the eight side chapels;
7. The four small horseshoe vaults above the corner chapels;
8. The slender columns in red marble.

Fig. 1. The dichotomy between the apparent formal structure and the actual load-bearing structure

In my studies of Guarini and in particular on the church of San Lorenzo, conducted as part of thesis work in architectural restoration,[1] I examined the composition of the double structure comprehensively by means of a systematic investigation. This work was divided into two overarching research threads:

a) The on-site collection of data inside the structure.

During this phase of investigation, that is, the onsite study of the structure, the notable complexity of the structure and the difficulties of access to internal technical spaces led to serious problems in reading the framework, making indispensable an ordering system for metaphorically deconstructing the system element by element, thus making it possible to classify each of them functionally and chronologically.

The two typologies of elements that were found most often are arches with vaulted structures (and their filling), which comprise the essential parts of the load-bearing structure, and the iron elements (tie-rods, keys, chains, bolts, etc.), which usually appertain to the interventions that are successive to the original construction.

The work included the compiling of forms relating to these two categories, with pertinent information listed, that were filled in directly on site, and which made it possible to organise the information relative to each element in the category. Three kinds of information were recorded on the forms:

– The first group served to *identify and situate* the element. A code on the form was also taped directly to the element, and contained information about the location within the structure;

– The second group contained information intended to provide a *detailed description of the element* according to the category it belonged to;

– The third group included a series of notes and relationships to other elements in the structure in order to *better understand the function* within the structure.

The three groups of information were backed up by sketches and drawings, including surveyed dimensions, photographs of whole elements, and photographs of details.

b) Archival and library research, and correlated research (such as studies of drawings, documents, decrees, journals, and Guarini's own writings), and a reorganisation of the categories and information.[2]

2 The nineteenth-century restoration

My purpose here is not to provide an exhaustive of the drawings attributed (and not) to Guarini, that is, the state of the art, which has already been addressed by Augusta Lange for the 1968 international conference entitled "G. Guarini e l'internazionalità del barocco" (where she attributed many drawings to Guarini, and many to collaborators). Rather, the object of this present paper is to examine the ten drawings of the double structure of San Lorenzo that are housed in the Archivio di Stato di Torino (sezioni riunite). These drawings were presented for the first time ever at the symposium "Guarino Guarini's Chapel of the Holy Shroud in Turin: Open Questions, Possible Solutions", 18-19 September 2006, which took place in the Archivio di Stato di Torino.

The drawings are all undated and unsigned, and come from what Augusta Lange called "His Majesty's Secret Archive" (*Archivio Segreto di Sua Maestà*), now known as "His Majesty's House" (*Casa di Sua Maestà*). They became property of the state on 1 January 1947.[3] Since that time they have remained stored in crates and only at the end of the 1980s were they catalogued in the inventory of the Archivio di Stato by Dr. Cecilia Laurora and her colleagues. As was noted in the thesis mentioned earlier, an examination of the crates of documents in the Archivio di Stato of Torino brought to light a series of drawings of the church of San Lorenzo, dating from 1795 to 1848 which, in addition to drawings of the facade, modification of the canon house and adjoining buildings, include a survey of the existing state of the building and structural drawings of the reinforcements made, which my thesis co-author and I dated to the period 1820-25 on the basis of other documentation.[4]

The drawings mentioned, archived as numbers 294-1/5 and 295-1/3, show the structure of the church in all its complexity and contain numerous notes about interventions that were necessary in order to reinforce the structure. They constitute the only survey in existence of the architectural structure except for that published in 1920 by the architects Proto and Denina.[5]

To date these drawings have been neither studied nor published. They are extremely valuable, in that they represent the various levels of the complex structure of San Lorenzo in its totality through plans, sections, and elevations which were drawing with notable graphic precision.

The drawings, part of the vast collection of documents entitled "Casa di Sua Maestà", are subdivided into three different series and comprise ten sheets, and all regard the double structure of San Lorenzo. They are presented for the first time in print here.

They are archived as follows:

- ASTO, AZIENDA REAL CASA (today "Casa di Sua Maestà"), Tipi e disegni, Torino, Chiesa di San Lorenzo, "piante della chiesa tutte senza data e autore". N. 294/1-5 sec. XVIII (Figs. 2, 3, 4, 5, 6).

- ASTO, AZIENDA REAL CASA (today "Casa di Sua Maestà"), Tipi e disegni, Torino, Chiesa di San Lorenzo, "piante e sezioni della chiesa tutte senza data e autore". N. 295/1-3 (figs. 7, 8, 9).

- ASTO, AZIENDA REAL CASA (today "Casa di Sua Maestà"), Tipi e disegni, Torino, Chiesa di San Lorenzo, "schizzo di arcate". Pencil drawing, N. 298 (fig. 10).

- ASTO, AZIENDA REAL CASA (today "Casa di Sua Maestà"), Tipi e disegni, Torino, Chiesa di San Lorenzo, "memorie di San Lorenzo". Schizzo a matita, non comprensibile o mancante di una parte. N. 299 (not reproduced here because it is not a drawing but is only a simple folder for holding drawings).

As the archival information shows, the drawings are unsigned and undated. The object of this present paper is to demonstrate both the attribution and the date.[6]

The drawings were shown at the symposium in a video, in order to be able to show the participants certain technical aspects by means of a comparison with the double structure in the church.

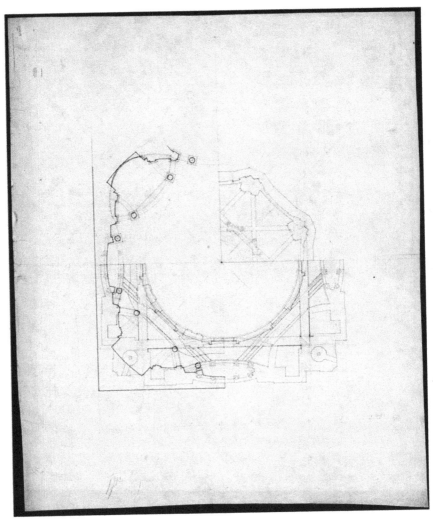

Fig. 2. Drawing 294/1

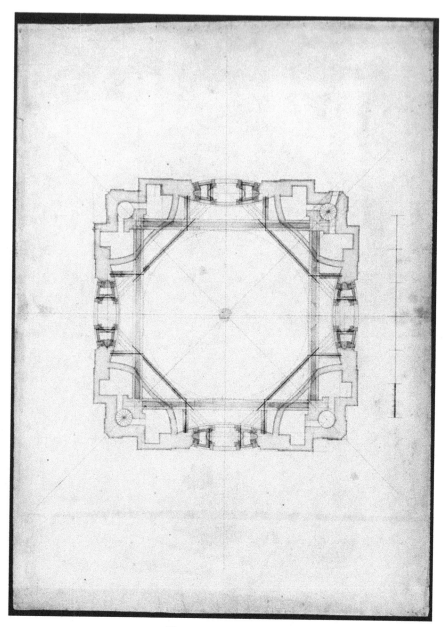

Fig. 3. Drawing 294/2

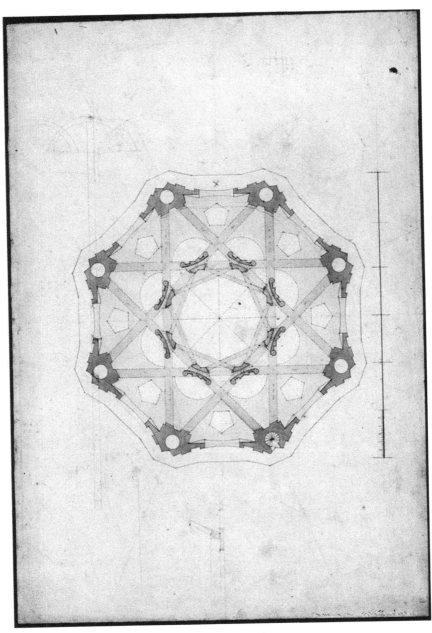

Fig. 4. Drawing 294/3

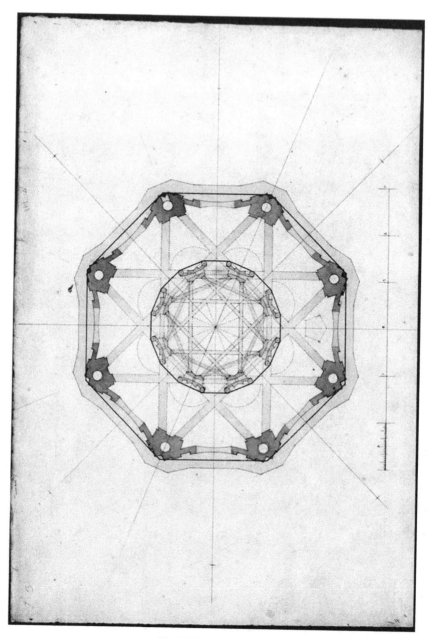

Fig. 5. Drawing 294/4

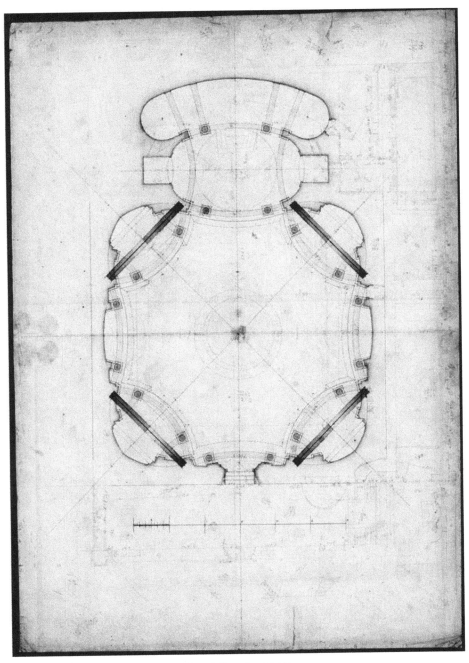

Fig. 6. Drawing 294/5

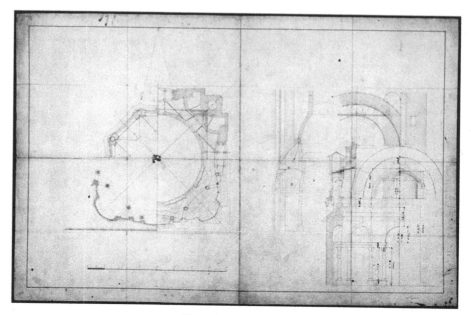

Fig. 7. Drawing 295/1

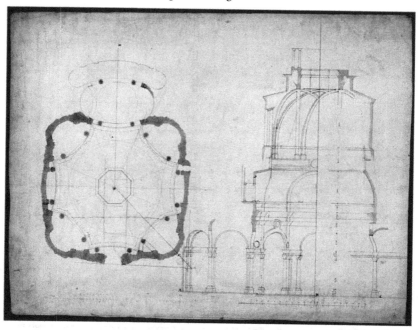

Fig. 8. Drawing 295/3

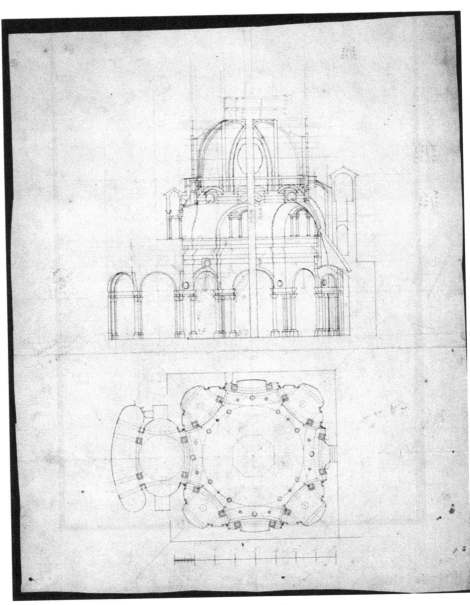

Fig. 9. Drawing 295/2

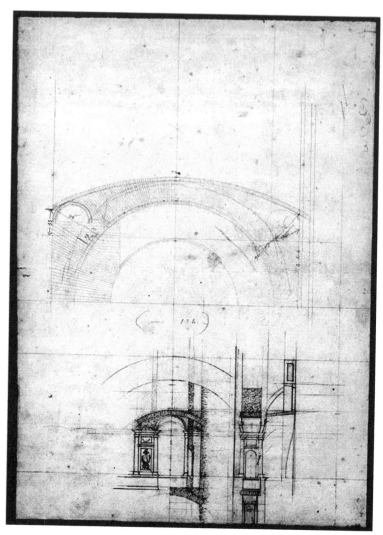

Fig. 10. Drawing 298

The drawings were analysed using two kinds of criteria: the paper used (sheet size, type of paper, type of grain, state of conservation, watermarks, the presence of holes, tears, attempts at restoration, and notes) and the drawing itself (subject, scale, line quality, background, notes, what appears on the back of the sheet, remarks).

The information gleaned was systematically recorded on forms, and made it possible to compare the drawings more easily. A first result was to prove that drawing no. 295/1, whose paper was reinforced by a thin hemp cloth glued on the back of it, served as a folder to contain drawings nos. 294/1-5. Further, as can be seen by the precision with which the drawings are laid out, from the completeness of the graphic information contained in them, and from the presence of squaring, drawing no. 295/1 is definitely the layout of the final project for the nineteenth-century restoration campaign. Drawings nos. 294/1-5 are thus preparatory drawings.

A comparison of the drawings to numerous related documents as well as explanatory notes by various artisans, has made it possible to find correspondences between the descriptions found other documents and the interventions of consolidation shown graphically on the drawings. This is further confirmed by correspondences between the letters used to indicate various elements in plan, elevation and sections in the drawings, and the letters used to refer to these views in the documents.

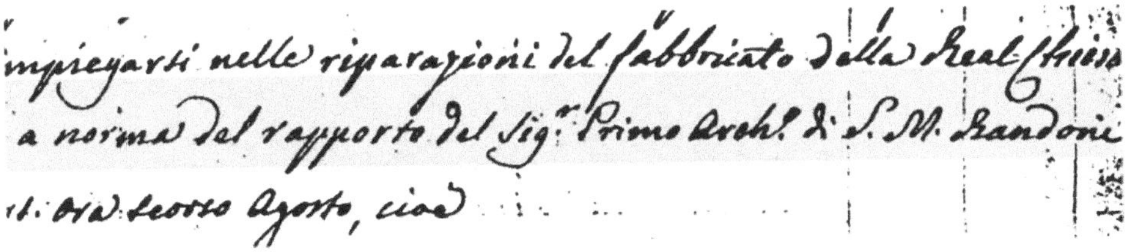
impiegarsi nelle riparazioni del fabbricato della Real Chiesa
a norma del rapporto del Sig.re Primo Arch.e di S. M. Randoni
31. ora scorso Agosto, cioè

Fig. 11

In fact, in the estimates provided by the master locksmith Tommaso Di Scalzo,[7] dated 28 October 1825, there are references to a report written by "His Majesty's chief architect Carlo Randoni (fig. 11): "calcolo istruttivo per la provvista della verosimile quantità di ferramenta necessaria impiegarsi nelle riparazioni del fabbricato della Real Chiesa di San Lorenzo a norma del rapporto del sig.re Primo Arch. Di S. M. Randoni in data de 31 ora scorso agosto". These contain references to the letters found on the drawings to identify specific elements, that is:

> *...n. 1 cerchio di simile lamone a diversi nodi ben teso da porsi esternamente alle basi delle colonne del cupolino e segnato nel profilo in istampa AB. ed espresso al capo 6° del rapporto in peso ----------------24.*
>
> *n. 4 tiranti doppi d'esso lamone per fasciare le radici segnate P. sul profilo, ed espresse al capo 3° del rapporto in peso -- 8* (figs. 12, 13).
>
> *n. 4 chiavi di lamone da uno la balla da mettersi in opera ai fianchi degli arconi principali segnati in pianta colle lettere GH. per impedire ulteriore dilatamento delle commessure nel mezzo degli arconi in peso --------------- ----------------74.*"[8] (figs. 14, 15, 16, 17, 18).

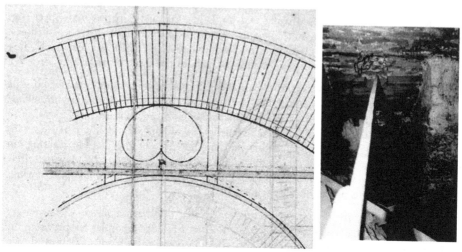

Fig. 12 (left). Detail of the elevation and section from drawing no. 295/1 showing a pre-existing wooden tie-rod by Guarini identified by the letter P at the intradox of the main arch (see fig. 7)

Fig. 13 (right). Tie-rod in iron at the intradox of the main arch that substituted the pre-existing wooden tie-rod by Guarini; note the larger opening in the masonry that had to be filled in left by the much larger wooden tie-rod that was removed, shown in drawing no. 295/1 (see fig. 12)

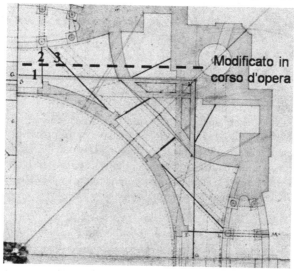

Fig. 14. Detail of the plan divided into quadrants shown in drawing no. 295/1 showing the section of the impost of the main arches with the insertion of various iron tie-rods marked with the letters GH. The dashed line indicates the actual position of the tie-rods, which is different from that planned due to changes as the work went forward (see figs. 16 and 18)

Fig. 15. Spiral stairway inside the corner tower where the vertical bar that acts as a hinge and connection between two tie-rods and the insertion of bolts to put them in tension can be seen, as shown in the detail of drawing no. 295/1 shown in fig. 14

Fig. 16 (left). Inside the double structure, at the extrados of the conical arches of the Serlian window; note that the iron tie-rod passes parallel to but does touch the main arch, as shown in drawing no. 295/1 (fig. 14)

Fig. 17 (right). Inside the double structure at the extrados of the spherical pendentives; note the wooden tie-rod of Guarini that still exists today, consolidated with metal pieces, as shown inthe detail of drawing no. 295/1 (fig. 14)

Fig. 18. Inside the double structure, at the extrados of the conical arches of the Serlian window; note the three tie-rods numbered 1, 2 and 3, arranged in triangular form, as shown in the detail of drawing no. 295/1 (fig. 14)

In drawing no. 294/5, which shows a plan of the church at the ground level, we can see indicated the insertion of a tie-rod and the construction of a masonry arch (fig. 19) that surmounted the altar of the corner chapel, serving to consolidate the intrados of the structure of the conical vaults (fig. 20).

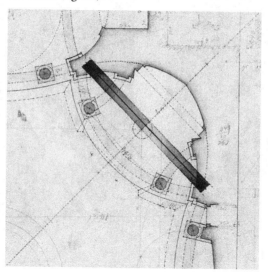

Fig. 19. Detail of drawing no. 294/5; note the indication of an arch to be constructed and the substitution of a tie-rod

Fig. 20. Inside the hidden space of the conical vaults; note the arch constructed to consolidate the intrados and the iron tie-rod

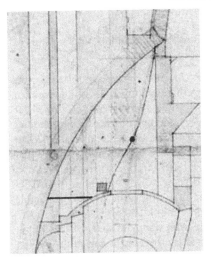

Fig. 21. Detail of drawing no. 295/1 showing where the conical vault intersects the pendentive, and showing the pre-existing wooden tie-rod of Guarini and the new iron tie-rod

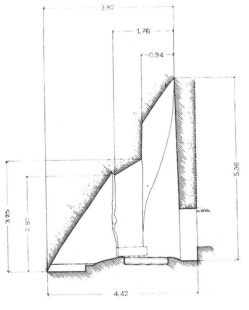

Fig. 22. Section by the author of the space inside the conical vault, part of the survey performed during work on the degree thesis completed under the guidance of Prof. Rosso in 1995

In the section of the corner altar with the conical vault above, shown in drawing no. 295/1 (fig. 21), it is possible to see, indicated in pencil, the insertion of the conical arch used to reinforce the intrados of the conical vault, as well as the new iron tie-rod and the pre-existing wooden tie-rod of Guarini.

In the surveys performed by architects Proto and Denina in 1920 we can see the same section shown in drawing no. 295/1, with the difference that the space of the conical vault is presumed by them to be a quarter-sphere, because they did not have access to the actual space.

The section shown in fig. 22 shows the survey performed by the author as part of thesis work done under the guidance of architect Franco Rosso. Here can be clearly seen the correspondence with the section shown in drawing no. 295/1 (fig. 21).

Thus, from the correspondences between the works described in the notes, contracts, and estimates, and the drawings analysed here, it is possible to attribute these drawings to architect Carlo Randoni.

Further confirmation of this statement is provided by drawings nos. 294/2, 3, 4 and 5, all of which are on the same kind of paper (laid paper with laid lines that average 29 mm) and feature fine lines in black and blue ink on wash backgrounds in pink, yellow and light grey. All of these characteristics appear in a drawing signed by architect Carlo Randoni and housed in the historic archives of the city of Torino[9], which refers to a different subject but is coeval, and is characterised, as I said, by the same kind of paper and graphic techniques.

Because of correspondences with the autograph drawings by Randoni, it is possible to attribute drawings nos. 294/2, 3, 4 and 5 to him as well.

In the pencil sketch designated as no. 298 (fig. 23) it can be seen that architect Randoni intended to fill in the side openings of the Serlian windows. This intention can be seen drawing in no. 294/2 as well (fig. 24). These interventions aimed at consolidation would be executed only in part: the two openings of the Serlian windows were not filled in, but the small vault to reinforce the lenticular structure over the Serlian windows was constructed. This reinforcement was necessary to stiffen the arches that receive the thrust of the diagonal arches (those marked as D in fig. 1); the intervention on the interior can be seen on the exterior facade where one of the windows is partially obstructed by the reinforcing vault passing in front of it (fig. 26). Inside the double structure, in the area above the Serlian window, a large moulded cornice (fig. 27) gives proof that the original structure of Guarini was visible from the main space of the church by means of other cornices which today are obstructed by the reinforcing vault (see fig. 25). Pencil drawing no. 298 can be attributed to architect Randoni because it is a preparatory drawing for the intervention to consolidate the Serlian window, shown in drawing no. 294/2, a working drawing we have already proven to be attributable to Randoni based on qualities of the paper and the drawing itself.

Fig. 23. Detail of drawing no. 298, a pencil sketch with views in elevation and section

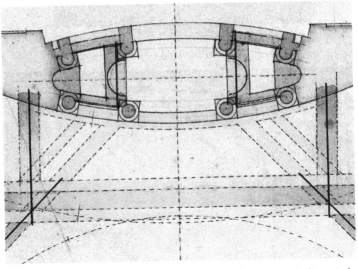

Fig. 24. Detail of drawing no. 294/2, showing the plan at the level of the Serlian window with the layout of the infill

Fig. 25. View from underneath of the small reinforcing vaults built in the lenticular structure of the Serlian window

Fig. 26. Photograph of the exterior facade of the space over the Serlian window; note the obstruction of the lower part by the reinforcing vault

Fig. 27. Inside the double structure (no longer visible from the main space of the church); note the moulded cornice, finished because it was exposed to view

Conclusions

The validity and success of a campaign of architectural restoration are directly proportional to the knowledge and comprehension of the construction science and techniques of the original project, and the stratification over time of other attempts at restoration. Given the complexity of Guarini's original concept for this church, the lack of original working drawings, and the particularities of the construction techniques used by this master architect, here it is especially difficult to intervene correctly and efficiently.

Over the course of successive centuries, Guarini's rich original design has been overlaid with additional structures and modifications intended to reinforce and consolidate the structure, making a reading of the original structure, already complex, even more difficult.

Synthesising all that we have learned about this church, we find ourselves dealing with an extremely articulated structure: Guarini created a shell that formally defines the interior space, but does not actually correspond with the load-bearing structure that supports the ribbed dome.

The complexity of the double structure, the visible and the hidden, could be considered as an exercise in technical virtuosity, an architectural contradiction created from the desire to fool the visitor. But in reality what Guarini created is a structural mechanism to achieve particular visual and compositional effects, aimed at representing a genuine mystical path that ascends to the Absolute, as well as an expression of Absolute through the geometric/mathematical structure of the architecture, whose profound nature is expressed dynamically through light, in space and over time.

One example of this is the emerging out of the darkness of a "Eternal Father" (see fig. 20), visible only the day of the spring equinox through an opening at the top of the vault over the corner altar dedicated to the Virgin, thanks to a studied and purposeful play of sunlight. Except for that one day, the fresco on the intrados of the conical vault is always hidden in darkness because it is inside the double structure.

Guarini's complex way of conceiving the load-bearing structures of stone, brick and wood is made evident by the unique double structure of San Lorenzo. Another double structure in stone and brick, also unique in its genre, is that of the Chapel of the Holy Shroud.

The extreme complexity of the construction of these buildings, as well as the great heterogeneity of the materials used, in comparison to modern materials such as reinforced concrete, render modern methods of analysis (such as the finite element method, or any of the other methods currently used in structural engineering) inadequate. Proof of this is the work done to consolidate the structure of San Lorenzo carried out by architect Randoni in 1825-1828 which, Randoni having immersed himself in Guarini's logic and through the workmanship of master masons, completely solved the structural problems.

The success of Randoni's work to consolidate the structure demonstrates that it is possible to intervene in Guarini's structures without the use of modern technology, especially when these prove to be altogether inadequate. Carlo Randoni's work is still intact today, without any apparent deterioration or cracking, proof that his working drawings – which show no signs of *pentimenti* or revision – are the fruit of thorough preparatory work, completely in harmony with Guarini's constructive logic.

Translation from the Italian by Kim Williams

Acknowledgments

The drawings here that are part of the Archivio di Stato di Torino (figs. 2-10) and the relative details are reproduced courtesy of the Ministero per i Beni e le Attività Culturali, Direzione Regionale per i Beni Culturali e Paesaggistici di Piemonte, prot. n. 5336/282800 of 5 August 2009.

I would particularly like to thank Dr. Cecilia Laurora of the Archivio di Stato di Torino for her helpfulness and collaboration; Franco Scalzo photographer of the Archivio di Stato di Torino; engineer Paolo Tarizzo for his valuable support for the presentation at the 2006 symposium on Guarino Guarini. Much of this material comes from my thesis, for which I thank Prof. Arch. Mario Dalla Costa, Arch. Paolo Fiora and Arch. Dario Lugato, my co-author on the thesis, Dr. Cristina Leoncini, and Prof. Arch. Franco Rosso for his collaboration in the survey of the space of the conical vaults.

For the kind permission to accede to the structure of the church, I thank the canons of San Lorenzo and especially the rector Don Franco Martinacci.

I also wish to remember my classmate, architect Beppe Demonte, who has since passed away, for the rendering of the "double structure".

Notes

1. Cristina Leoncini, Ugo Quarello, "*La doppia struttura della chiesa di S. Lorenzo del Guarini, esempio di architettura nella Torino del seicento*", thesis advisor Prof. Arch. Mario Dalla Costa, Politecnico di Torino, faculty of architecture, academic year 1994/95.
2. For a complete description, see "La doppia struttura..." *op cit.*, pp. 21, 22, 23.

3. As Lange wrote, "I believe instead that the archive of papers of the Crown or His Majesty's Secret Archive, which became property of the state on 1 January 1947, still housed in the Palazzo Reale in Torino, and to be precise as part of the categories of *Designs for buildings* and *Plans of the holdings of the Savoy House in Piedmont*, must contain, along with the graphic documentation of the building activities that we are totally lacking regarding the ducal and royal palaces, other drawings relative to the Chapel of the Holy Shroud and S. Lorenzo" (*Ritengo invece che il fondo di carte della Corona o Archivio Segreto di S. M., divenuto proprietà demaniale nel 1 gennaio 1947, tuttora conservato nel Palazzo Reale di Torino e precisamente nelle categorie progetto di fabbricati e Planimetrie dei beni di Casa Savoia in Piemonte, non possa non contenere insieme con quella documentazione grafica delle vicende costruttive che ci manca assolutamente dei palazzi ducali e reali, altri disegni relativi alla S. Sindone e a S. Lorenzo*). Cf. Augusta Lange, *Disegni e documenti di G. Guarini*, stà in *Guarino Guarini e l'internazionalità del barocco*, Atti del Convegno, vol. 1, pag 92, Accademia delle Scienze, Torino 1970.

4. Cf. "La doppia struttura..." *op cit*, p. 99.

5. L. Denina, A. Proto, La real chiesa di San Lorenzo a Torino, in *L'architettura Italiana*, vol. xv, no. III, Crudo, Torino, 1920, pp, 34-38. Denina and Proto did not know about these drawings, as is clear from the section of the "conical vaults", which they presumed to be a quarter-sphere.

6. In 1995, in my degree thesis, I had attributed some drawings to the architect Carlo Randoni, (cf. pp. 104/105); at the 2006 symposium on Guarini in Torino, during the part of my presentation given in the hall of the Archivio di Stato where the drawings were displayed, I declared the certain attribution to Carlo Randoni of other drawings in the collection. From the date of the symposium, 19 September 2006, these drawings are still archived with the same nomenclature cited above. However, these same drawings are cited in a note in the essay Isabella Massabò Ricci entitled "I disegni di Guarino Guarini nell'Archivio di Stato di Torino. Alcune questioni di metodo", published in December 2006 in the book *Guarino Guarini* edited by G. Dardanello, S. Klaiber, H. A. Millon (Allemandi, Torino) and are described thus: nota 22, "Il vasto fondo documentario...(ASTO, Tipi e disegni Casa di S.M., Disegni San Lorenzo, n. 294/1; Carlo Randoni e studio, "Pianta della Chiesa di San Lorenzo sezionata a quattro livelli differenti", 1825/1828; n.294/2, Carlo Randoni e studio, "Pianta a livello dell'imposta della Cupola di San Lorenzo")"; cf. note 22, p. 7.

7. ASTO, Casa di Sua Maestà, periodo 1817-1870 Vol. V, 1825-1826, pp. 191, 192.

8. ASTO, Casa di Sua Maestà, periodo 1817-1870 Vol. V, 1825-1826, pp. 191, 192.

9. Archivio storico città di Torino (ASCT), progetti edilizi, I° categoria, 1825/14 (2 drawings, plan and elevation).

About the author

Ugo Quarello was born in Torino in 1968 and earned his degree in architecture at the Politecnico di Torino, with a thesis on the double structure of San Lorenzo by Guarini. He worked with architect Leonardo Mosso at the Alvar Aalto Institute in Pino Torinese. He has undertaken various studies, including research on the medieval village of Cervo Ligure, analyses of Aalto's unbuilt design for the Quartiere Patrizia in San Lanfranco, Pavia, feasibility studies for the hillside areas of Torino, preliminary studies for a museum of ecology near Superga. He has been a practicing architect since 1997, developing his personal artistic creativity and emphasizing an interdisciplinary approach that unites research, history, landscape and architectural design.

Patricia Radelet-de Grave

Université catholique de Louvain
Edition Bernoulli
2 chemin du cyclotron
B-1348 Louvain-la-Neuve, BELGIQUE
Patricia.Radelet@uclouvain.be

Keywords : Guarino Guarini, astronomie, cosmologie

Research

Guarini et la structure de l'Univers

Abstract. While Guarino Guarini is well known as an architect, his intellectual work was not limited to architecture, and three of his publications concern astronomy. This present paper concentrates on the first part of the 1683 *Coelestis mathematicae*. It appears clear that Guarini refused to take any official position in defence of either heliocentricity or geocentricity.

L'œuvre publiée de Guarino Guarini

Guarino Guarini est bien connu en tant qu'architecte et plusieurs de ses réalisations, parmi les plus belles ornent la ville de Turin. Pourtant l'œuvre intellectuelle de Guarini, ne se limite pas à l'architecture comme le montre la liste de ses publications où nous avons marqué d'un astérisque les trois publications relatives à l'astronomie.

Placita philosophica, Parisiis, apud Dionysium Thierry, 1665

Trattato di fortificatione, che ora si usa in Fiandra, Francia, & Italia, Torino, Per gl'Eredi Gianelli, 1666

Euclides adauctus et methodicus mathematicaque universalis, Turin, Zapata, 1671 réédit 1676

Modo di misurare le fabbriche, Torino, per gl'heredi Gianelli, 1674

* *Compendio della sfera celeste,* Torino, appresso Giorgio Colonna, 1675

* *Leges temporum et planetarum,* Torino, Per gl'Eredi Gianelli, 1678

* *Coelestis mathematicae pars prima et secunda,* Mediolani, ex Typographia Ludouici Montiæ, 1683

Disegni di architettura civile ed ecclesiastica, Torino, Per gl'Eredi Gianelli, 1686

Novum Theatrum Pedemontii et Sabaudiae..., Hagae-Comitum: sumptibus et cura R. C. Alberts, 1726

Architettura civile, Torino, G. Mairesse all'insegna di Santa Teresa di Gesù, 1737

Fig. 1. Portrait de Guarino Guarini

Nexus Network Journal 11 (2009) 393–414
NEXUS NETWORK JOURNAL – VOL.11, No. 3, 2009 **393**
DOI 10.1007/s00004-009-0005-9; *published online* 5 November 2009

Le premier de ces travaux rassemble ses cours de physique et de philosophie, selon certains, Guarini y tente une réforme de l'Aristotélisme. Cinq travaux, soit la moitié de son œuvre publiée, concernent, on ne s'en étonnera pas, l'architecture et les fortifications. On y trouve encore une édition commentée d'Euclide et trois textes, ce qui est une proportion considérable pour une œuvre qui ne compte que dix titres, consacrés à l'étude de la cosmologie. La somme de travail, de lecture de textes anciens, de mesures et d'observations contenue dans ces travaux prouve indéniablement l'importance que Guarini attribuait à cette étude. Nous nous concentrerons sur la première partie des *Coelestis mathematicae* car la deuxième partie est un traité de gnomonique, que les *Leges temporum et planetarum,* contiennent principalement des tables de mesures et que je n'ai malheureusement pas vu le *Compendio della sfera celeste.*

Mais avant cela quelques remarques plus générales sur les opinions cosmologiques à l'époque de Guarino Guarini.

Le contexte de l'étude de la cosmologie

L'histoire des sciences se contente trop souvent de quelques dates phares pour jalonner l'histoire de la cosmologie.

1543, Copernic, *de Revolutionibus,*

1585, Stevin, *Whisconstige Ghedachtenissen,*

1596, Kepler *Mysterium cosmographicum,*

1609, Kepler *Astronomia Nova*

1610, Tycho Brahe *Astronomiae instauratae progymnasmata.*

1619, Kepler, *Harmonices mundi*

1622, Longomontanus, *Astronomia Danica*

1632, Galilée, *Dialogo*

Il semble alors qu'en 1632 tout ait été dit et que l'intérêt d'un texte publié en 1683 semble dès lors minime.

Mais on oublie, en faisant cela que les religieux catholiques, dont Guarino Guarini faisait partie puisqu'il appartenait à l'ordre des Théatins, et qui constituent une grande partie de l'élite intellectuelle, sont muselés par la mise à l'Index du livre de Copernic en 1616. Les Protestants rencontrent également des difficultés en défendant Copernic. Il faut donc être prudent face aux opinions publiées par les auteurs et ne pas prendre pour argent comptant le fait que l'un ou l'autre semble défendre le système de Ptolémée. Souvent leur opinion secrète n'apparaît pas dans les publications.

Opposition des images et du contenu

Prenons l'exemple de Jan Ciermans, Jésuite, élève de Grégoire de Saint-Vincent. Ciermans publie en 1640, un cours à l'intention des élèves des collèges jésuites.

Bien qu'il cite tant Copernic, que Kepler et Van Lansberg et Wendelen, les plus ardents défenseurs belges de Copernic, Ciermans ne prend pas officiellement position. Il présente les théories de ces auteurs comme diverses hypothèses toutes envisageables, ce qui était autorisé par l'Eglise. Il va jusqu'à envisager d'autres hypothèses :

Fig. 2 et 3. Vignettes illustrant les *Disciplinae mathematicae* publiées par Jan Ciermans en 1640

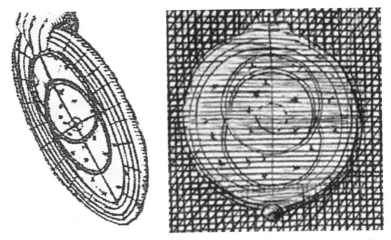

Fig. 4 et 5. Détails de la Fig. 1 et de la Fig. 2, reproduisant la figure donnée par Kepler pour illustrer sa découverte de la forme elliptique de l'orbite de Mars (Fig. 6)

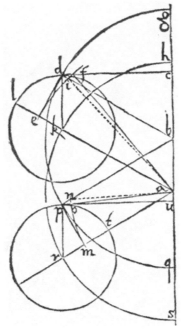

Fig. 6. Figure donnée par Kepler dans *l'Astronomia Nova* pour illustrer sa découverte de la forme elliptique de l'orbite de Mars

Que le Soleil soit fixe et que la Terre se meuve, de là on peut montrer exactement, que dans ces hypothèses, on sauve tous les phénomènes, et on en conclut les mêmes choses pour la Lune, on peut en effet trouver un système où tout aussi bien cette dernière, comme le Soleil chez Copernic ou la Terre chez Tycho, reste immobile.

Ciermans estime que l'on peut choisir de placer le centre fixe du système planétaire n'importe où, on pourra toujours sauver les phénomènes. Mais il s'agit là d'hypothèses. Ciermans n'aborde pas l'éventualité de la réalité de ces situations.

Si nous analysons attentivement les jolies vignettes qui illustrent le cours (Fig. 2 et 3), nous y découvrons deux figures (Fig. 4 et 5) copiant celle de Kepler (Fig. 6) lorsqu'il explique sa découverte de la forme elliptique de l'orbite de Mars dans l'*Astronomia Nova* en 1609.

Ciermans ne prend donc pas officiellement position quant à la réalité du système héliocentrique, mais une vignette illustrant son cours de 1640 montre deux *Putti* dont l'un regarde les phases de Vénus au moyen d'une lunette alors que l'autre se cache les yeux pour ne pas voir la réalité (Fig. 7). Nous verrons avec Guarini, l'importance de l'observation des phases de Vénus.

Fig. 7. Vignette illustrant les *Disciplinae mathematicae* publiées par Jan Ciermans en 1640

Le plan des traités d'astronomie

L'influence de l'introduction d'Ossiander au *de Revolutionibus* de Copernic sur un grand nombre de traité est remarquable. Rappelons que dans cette introduction, Ossiander, par peur de la réaction prévisible au bouleversement proposé par Copernic présente ces idées comme une hypothèse, non réelle mais facilitant les calculs. Beaucoup d'auteurs vont saisir cette opportunité et présenter les différents systèmes sur pied d'égalité et sans jugement quant à leur réalité. Cette méthode permettait de masquer son opinion. C'était aussi la manière de faire qui avait été exigée de Galilée par Urbain VIII pour obtenir de l'Eglise la permission de rédiger le *Dialogo*. Il convenait évidemment dans ces circonstances de masquer réellement ses opinions. Ce que Galilée ne fit pas. Pas plus que ne le fait un Stevin, qui pourtant adopte le plan de mise en parallèle des différents systèmes, mais y sélectionne un système comme correspondant à la réalité, celui de Copernic. Comparons son plan à celui de Giovanni Batista Riccioli (1598-1671) qui assume des responsabilités importantes au sein de la Compagnie de Jésus. Son plan est tiré de l'*Almagestum Novum* publié en 1651.

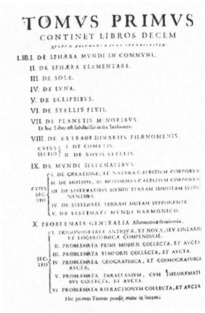

Fig. 8 et 9. Le frontispice de l'*Almagestum Novum* de 1651 et le plan des dix premiers livres

Nous y lisons à propos du

IX *système du monde* :　I. *De la création et de la nature des corps célestes*

II. *Des mouvements et des moteurs des corps célestes*

III. *Du système du monde supposant la Terre immobile*

IV. *Du système supposant la Terre mobile*

V. *Du système harmonique du monde*

Les hypothèses d'une terre mobile ou fixe sont donc apparemment mises sur le même pied.

Chez Stevin, le plan de l'*Astronomie*, IIIe partie de la *Cosmographie* n'est pas très différent.

Livre I *De l'invention du cours des Planètes, et des estoiles fixes, par les Ephémérides observées, le tout fondé sur la supposition que le terre est stable ou fixe; c'est en un mot , sur l'hypothese de terre immobile.*

Livre II *De l'invention du cours des Planètes, par voye Mathematique, avec l'hypothèse de terre immobile et de la première inegalité*

Livre III. *de la seconde inégalité où se trouve l'hypothèse de terre mobile de Copernique*

Le Père Claude François Milliet Dechales, choisit aussi de mettre les différents systèmes en parallèle comme le montre la table des matières de son *Astronomia,* Traité XXVIII de son *Cursus seu Mundus mathematicus* de 1690.

Fig. 10. Plan de la partie astronomique du *Cursus seu Mundus mathematicus* publié par le Père Claude François Dechales en 1690

On peut y lire :

    Proponitur systema Ptolemaicum terrae immotae

    Egyptiacum systema terrae immotae

    Systema tychonicum terrae immotae

    Systemat copernicanum terrae motae

Certains auteurs comme Ciermans ne se prononcent que par l'image. D'autres comme Stevin prennent clairement position en faveur du système de Copernic. Mais ils sont rares. Riccioli opte officiellement pour le système hybride de Tycho, où toutes les planètes tournent autour du Soleil qui lui tourne autour de la Terre, les entraînant toutes dans ce mouvement. Pourtant Riccioli concédait que le système de Copernic était le plus pratique mais non réel. On peut d'ailleurs se demander si son frontispice ne montre pas, grâce à une astuce de perspective, sa prédilection pour le système de Copernic. Guarino Guarini n'échappe pas à la règle et présente les trois systèmes de manière équivalente.

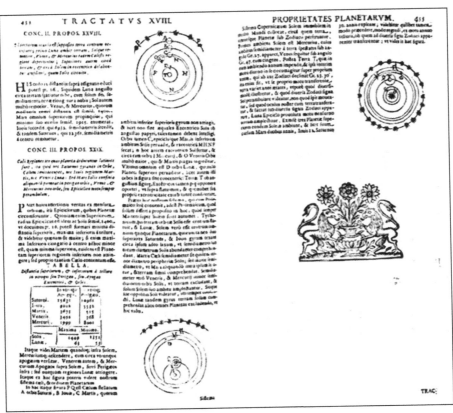

Fig. 11. Pages 454-455 extraites du *Coelestis mathematicae* publié par Guarino Guarini en 1683

Les *Coelestis mathematicae* publié par Guarino Guarini en 1683

Le plan du Traité et description générale

Tractatus XI : Eclipsis lunae demonstrata

Tractatus XII : Eclipsis solis ostensa

Tractatus XIII : Eclipsis terrae deprehensa

Tractatus XIV : Trium planetarum superiorum hypoteses stabilitae

Tractatus XV : Duorum inferiorum hypoteses stabilitae

Tractatus XVI : Latitudo minorum planetarum dimensa

Tractatus XVII : De passionibus planetarum

Tractatus XVIII : Proprietates planetarum expositae

Tractatus XIX : Stellae fixae indigitatae

Tractatus XX : Asterismorum figurae descriptae

Tractatus XXI : De fixis singulis caelo visibilibus

Un rapide tour d'horizon de la table des matières de cet ouvrage, confirme l'importance accordée par Guarini aux mesures, importance qui nous avait déjà été montrée par les nombreuses tables auxquelles il consacre plusieurs ouvrages. Le fait de rester près des mesures entraîne d'office une allure ptoléméenne car ne l'oublions pas jusqu'à nos sondes spatiales, toutes les mesures ont toujours été géocentriques. Mais est-ce tout ?

Les auteurs cités par Guarini

Inventorions et analysons rapidement l'avis sur le système à adopter, de ceux qui sont cités par Guarini.

Excluons d'abord de cette analyse les personnages incontournables de ce sujet, comme Copernic, Stevin, Kepler, Tycho Brahe et Galilée qui sont mentionnés par Guarini et dont les opinions sont bien connues. Nous reviendrons sur les positions de Guarini vis-à-vis de ces personnages dans notre analyse de ses *coelestis mathematicae*.

Excluons aussi ceux à qui Guarini fait appel parce qu'ils donnent des mesures dont il a besoin lors de l'établissement de ses tables mais aussi lors de ses raisonnements théoriques. Les opinions de ces auteurs sur le système héliocentrique ne nous apprend rien sur le point de vue de Guarini.

Mais Guarini cite bien d'autres auteurs dont voici la liste

Christiaan Huygens (1629 - 1695)

Ismaïl Bouillaud (1605 – 1694)

Claude Milliet Dechasles (1621 – 1678),

Caramuel y Lebkovich, (1606 – 1682)

Godefroid Wendelen (1580 – 1667)

Dominique Cassini (1625 – 1712)

Christian Longomontanus (1562 – 1647)

Francesco Fontana (1602 – 1656)

Philipe van Lansberg (1561 – 1632)

Nicolas Zucchi (1586 - 1670)

Gérard Jean Vossius (1577-1649) ou Isaac Vossius (1618-1688)

Force est de constater que les opposants au copernicnisme sont absents et que les défenseurs sont largement majoritaires par rapport aux abstentionnistes. Mais on ne peut pas tirer de conclusion définitive d'une telle constatation.

Revenons sur certains chapitres des *Coelestis mathematicae*

Nous voudrions dans ce qui suit montrer la difficulté qu'il y a, en lisant attentivement le texte , à décider de l'opinion profonde de Guarini. Il est toujours possible de trouver quelques phrases qui semblent définitive en les sortant de leur contexte. Mais immédiatement d'autres se présentent qui pourraient très bien soutenir le contraire. Notre but n'est donc pas de démontrer que Guarini a adopté l'une ou l'autre position mais plutôt de montrer à quel point il a laissé planer le doute.

Le livre s'ouvre sur un intitulé des plus classiques

Tractatus I: *Sphoera caelestis descripta* [*Coelestis mathematicae*, Vol. 1, p. 1]

(Traité I : Description de la sphère céleste.)

Guarini semble se placer d'entrée de jeu dans la lignée géocentrique du *de sphaera* de Sacrobosco,

Expensio 1 : *De caeli rotunditate.*

(Premier développement : De la sphéricité du ciel.)

Theor I. Propos I. : *Coeli motus circularis est quoad sensum.*

(Théorème I. Proposition I : Le mouvement du Ciel est circulaire pour ce qui concerne les sens.)

Et le théorème I, Proposition I sur le mouvement circulaire du ciel nous conforte dans cette opinion. Mais que devons-nous penser de la restriction : pour ce qui concerne nos sens. Le mouvement circulaire du Ciel serait-il dû à une erreur de nos sens ? Cette remarque peut recouvrir le principe de relativité optique, c'est-à-dire le fait que c'est la rotation de la Terre qui nous fait croire au mouvement du Ciel. Guarini qui a lu Galilée ne peut manquer de le savoir. Ebranlé déjà par cette remarque que dire de l'allusion aux orbites elliptiques de Kepler qui suit immédiatement.

… Neque obstat quod planetae juxta Ptolemaeum per excentricos delabantur, aut juxta Keplerum per ellipses.

(… Personne ne s'oppose à ce que les planètes suivant Ptolémée parcourent des excentriques ou suivant Kepler des ellipses.)

Les deux systèmes sont mis sur le même pied. Personne ne s'y oppose parce que l'ellipse ne diffère pas fortement du cercle.

… excentrici vere circuli sint et ellipses, si tamen admittendae, non valde discrepant a criculo.

(... Les excentriques sont de véritables cercles et les ellipses si on les admet, ne diffèrent pas beaucoup du cercle.)

Nous devons pourtant être attentifs et ne pas conclure trop vite. Comme chez Ciermans, et contrairement à nos habitudes, les orbites elliptiques sont admises facilement et ne sont pas dans l'esprit des gens de l'époque synonymes d'héliocentrisme.

Les traités II à VII

Ces traités traitent de la mesure. Comment faire des mesures ? Quelle est l'importance du temps, temps civil et temps astronomique, puis les difficultés de mesures, la parallaxe, la réfraction.

Tractatus VII : Observationes astrorum operi traditae.

(Traité VII : Observations des astres transmises dans les œuvres.)

Après avoir inventorié les difficultés des mesures, Guarini donne une liste, tirée d'auteurs anciens, des irrégularités, du Soleil, de la lune, des trois planètes supérieures, des planètes inférieures. En montrant les difficultés liées à la mesure, Guarini nous a en fait mis en garde contre les résultats qu'il donne à présent. Ce chapitre se termine par une phrase étrange :

Trademus autem omnium irregularitatum quantitatem, tum Longitudinis, tum Latitudinis, & modum quo quantitates singulae decernuntur cum nostro motus explicabimus infra [Coelestis mathematicae, Vol. 1, p. 168].

(Nous traiterons de la grandeur de toutes ces irrégularités, tant en longitude qu'en latitude, et de la manière dont chacune de ces irrégularités peut être tranchée lorsque nous expliquerons, plus loin, notre mouvement.)

Guarini va-t-il à son tour nous proposer un système planétaire ? Mais avant cela, il va rendre compte des systèmes proposés jusqu'ici.

Tractatus VIII : Theoriae planetarum descriptae.

(Traité VIII : Description de la théorie des planètes.)

Ce chapitre est introduit de la manière suivante qui résume ce qu'il vient de faire. Semer le doute sur les résultats de mesure et dès lors discréditer en bloc toutes les lignes tortueuses qui ont été inventées pour décrire le mouvement des planètes.

Cum, ut vidimus, ex observationibus colligatur motus caelestes esse inaequales, Astronomis ea cura fuit tales circulos, linearumque amphractus invenire, per quos ducti Planetae illas inaequalitates exprimerent. Inter vero alias hypotheses, quibus sub lege errores astrorum rediguntur Ptolemaica adeo posteritati placuit, & concinavisa est, ut nemo eam novis circulorum ambagibus temerare, usque ad Copernicum ausus fuerit. Sed is audacior multa mole motum telluris astruens novas caelo tricas invexit, quem Keplerus, Bullialdusque, aliisque permulti secuti insuper, & pro circulis substituerunt Ellipses, putantes melius errorum caelestium incertitudines posse ellipticis gyris rapraesentari ; quamvis, et spes eos fefellerit. Quapropter nedum antiqua siderum Theoria docenda est tanquam basis, fundamentumque caetererum, captuque facilior ; sed & alia proponendae, ut plenam lector omnium modorum, quibus motu caelestis prostaphereses aequantur, obtineat cognitionem [*Coelestis mathematicae*, Vol. 1, p. 169].

(Puisque, comme nous l'avons vu par l'ensemble des observations rassemblées, les mouvements célestes sont irréguliers, les astronomes ont inventé pour y remédier, des cercles et des lignes tortueuses (amphractus), au moyen desquelles, ils expriment les irrégularités du mouvement des Planètes. Parmi les différentes hypothèses, qui furent rédigées pour expliquer les irrégularités des Astres, celles de Ptolémée plurent beaucoup à la postérité et furent jugées agréables, au point que personne n'osa proposer d'autres sinuosités des cercles jusqu'à ce que Copernic ne l'ose. Mais cet audacieux, inventât de nouveaux mouvements du ciel et surtout le mouvement de la Terre et fut suivi par Kepler, Bouillaud et beaucoup d'autres, qui de plus, remplacèrent les cercles par des ellipses, pensant que ces tours elliptiques rendraient mieux compte des erreurs célestes, mais leurs espoirs furent déçus. C'est pourquoi, il convient d'enseigner l'ancienne théorie en tant que base et fondement des autres, et plus simple à comprendre ; mais aussi les autres pour permettre au lecteur de connaître tous les moyens de ramener les mouvements des cieux à la prostaphérèse.)

Bien que la théorie de l'audacieux Copernic semble attractive pour Guarini, il n'en ressort pas moins qu'elle ne rend pas mieux compte des « erreurs célestes ». Dès lors, quelle que soit l'opinion que l'on puisse avoir personnellement, Guarini conseille de considérer et d'enseigner l'ensemble des systèmes proposés. Nous sommes arrivé à la conclusion de tous les auteurs mentionnés plus haut dont Ciermans. Guarini ajoute que de cette manière on apprendra à mieux connaître la *prostaphérès*. Il explique la nécessité de ce calcul engendré par le fait que le centre du mouvement vrai n'est pas le centre du Monde.

Pars vero, quae additur vocatur Æquatio, seu Prostapheresis, & Motus in excentrico vocatur Anomalia, seu Irregularitas, non quia motus in ipso sit inaequalis ; sed quia causat inaequalitatem motus veri, ob eius centrum extra centrum Mundi collocatum [*Coelestis mathematicae*, Vol. 1, p. 170].

(La partie ajoutée est appelée équation, ou prostapheresis, et le mouvement dans l'excentrique est appelé anomalie ou irrégularité, non parce que le mouvement y est inégal mais parce que le fait que son centre ne coïncide pas avec le centre du monde cause une inégalité du mouvement vrai.)

La non identité, du centre du monde avec le centre de l'orbite entraîne un terme supplémentaire dans l'équation de l'orbite. Certains appellent prostaphérèse l'opération qui ramène le produit de deux sinus à une somme de deux sinus.

Ensuite Guarini va inventorier tout ce que l'on a trouvé pour palier à ces « erreurs célestes » C'est-à-dire pour mieux rendre compte de la forme des orbites ou encore comme le disaient les anciens pour mieux sauver les phénomènes. Il va ainsi énumérer les outils de la théorie ptoléméienne, les excentriques, épicycles et autres déférents.

Expensio I : De excentrico

(Expensio I : De l'excentrique.)

Expensio II : *De duplici excentrico*

Quoniam in aliquibus Planetis, immo in omnibus secundum aliquos unicus excentricus irregularitatem eorum non eliminat ; hinc Astronomi duplicem Excentricum protulerunt [*Coelestis mathematicae*, Vol. 1, p. 172].

(Expensio II : Du double excentrique

Comme pour certaines planètes, si pas pour toutes, toutes les irrégularités ne disparaissent pas avec un excentrique, les astronomes proposèrent le double excentrique.)

Expensio III : *de Epicyclo*

Epiciclus est parvus circulus, cuius centrum defertur per circumferentiam alterius circuli. Inventusque est ad salvandam aliam irregularitatem Planetarum, quam vocant respectivam ob connexionem quandam, quam gerit cum Sole, Superiorum quidem qui circa punctum motus medij, ut Sol circa terram moventur eodem tempore circulum cum Sole absolventes, inferiorem autem, qui se girant circa Solem eum pro centro motuum suorum, recognoscendo [*Coelestis mathematicae*, Vol. 1, p. 175].

(Expensio III : de l'Epicycle

L'épicycle est un petit cercle dont le centre est déplacé par la circonférence d'un autre cercle. Il est inventé pour sauver d'autres irrégularités des Planètes que l'on appelle respectives à cause d'une connexion qu'elles ont avec le Soleil, les supérieures elles mêmes qui tournent autour d'un point, d'un mouvement moyen, comme le Soleil est mû autour de la Terre dans le même temps que le Soleil autour de la Terre et les inférieures aussi qui tournent autour du Soleil en le reconnaissant comme centre de leur mouvement.)

Expensio IV : *De epiciclo in excentricis se movente*

(Expensio IV : de l'Epicycle se mouvant sur un excentrique)

Expensio V : *De duplici Epicyclo*

(Expensio V : Du double Epicycle)

Ce double épicycle a été proposé par Tycho et Longomontanus pour expliquer les irrégularités difficiles de la Lune.

Après avoir inventorier tous ces remèdes inventés pour rectifier les irrégularités des mouvements célestes depuis l'Antiquité, Guarini poursuit sans transition en introduisant les ellipses, et en les classant elles aussi parmi ces remèdes.

Expensio VI : *De Ellipsi*

Keplerus, sed foelicius Bullialdus Ellipsim pro moderamine primae irregularitatis coelis inveserunt, & licet illi absurdum motum terrae prosupponantur, potuissent tamen ipsum adhibere, & forte facilius, quamvis tellus non agitarent ad salvandos caelestes lapsus [*Coelestis mathematicae*, Vol. 1, p. 183].

(Expensio VI : de l'Ellipse

Kepler, et mieux encore Bouillaud, introduirent l'ellipse pour diriger les premières irrégularités du ciel et laissèrent présupposer que cet absurde mouvement de la Terre, puisse s'appliquer, et beaucoup plus facilement qu'une Terre qui ne s'agiterait pas, à sauver les glissements du ciel.)

Notons ici l'absurdité du mouvement de la terre ou encore de son agitation. Des mots qui, tirés de leur contexte, pourraient laisser croire à l'avis classique et ptoléméen de Guarini, mais ce serait oublier sa remarque initiale sur les sens qui pourraient nous abuser.

Expensio VII : *Motuum caelestium ambages circulis, seu ellipsibus expressae*

Ex observationibus vidimus motus caelestes regulares non esse, aut aequales solemque annuam suam irregularitatem obtinere caeteros Planetas omnes duas consequi, alteram absolutam qua redirent fere post certum tempus ad eandem partem zodiaci, alteram respectivam, que motum Solis respicit, ita quod finiret cum Sol ipsis accederet, iisque coniungeretur. Hic vero volumus plenius declare quomodo hi motus per circulos, seu Ellipses ab Astronomis sal ventur generaliter tamen, & prescindendo a particularibus cuiuscunque Planete orbibus, & hoc ut illorum usus innotescat [*Coelestis mathematicae*, Vol. 1, p. 186].

(Expensio VII : Mouvements célestes exprimés par deux cercles, c'est-à-dire une ellipse

Les observations nous ont fait voir que les mouvements du ciel n'étaient pas réguliers ni égaux, et que l'irrégularité annuelle du Soleil a deux conséquences sur les autres Planètes, l'une absolue, qu'elles reviennent après un certain temps à la même place dans le Zodiac, l'autre respective, qu'elles respectent le mouvement du Soleil de manière à finir [leur mouvement] lorsque le Soleil termine [le sien], comme conjugué à lui. Ici nous voulons expliquer pleinement comment les Astronomes sauvent ces mouvements d'abord de manière générales et puis en déterminant pour une Planète quelconque particulière son orbite et tout ce qui est nécessaire pour l'utiliser.)

Cette constatation du mouvement coordonné du Soleil et des planètes avait été faite depuis longtemps et avait justifié le rôle royal attribué au Soleil qui règne sur les planètes qui l'accompagnent. Mais il est évident pour celui qui a compris les choses que ce comportement est une conséquence immédiate de l'héliocentrisme.

Corollarium I

Hinc infere licet, ut diximus etiam supra, loco Aequantis, & Deferentis posse substitui Ellipsim, vel loco aequantis Excentricum, & loco Deferentis, & Epicicli duplex Epiciclus, vel etiam secundum alios eleminari potest Aequans, & solo Deferente uti pro prima irregularitate absoluta, pro secunda Epiciclo, vel etiam alio Excentrico ; essetque morosum velle omnes modos, quibus Planetae motus horum circulorum varia compositione potest salvari, velle prolixius explicare [*Coelestis mathematicae*, Vol. 1, p. 188].

(Corollaire I

De là on peut inférer, comme nous l'avons déjà dit plus haut, que l'on peut substituer des ellipses au lieu des équants et déférents, ou à la place des excentriques et à la place des déférents et des épicycles ou double épicycle ou encore suivant d'autres que l'on peut supprimer les équants et ne garder que les déférents pour la première irrégularité absolue; et pour la seconde un épicycle ou bien un autre excentrique ; et il est difficile de connaître toutes les manières dont les Planètes peuvent sauver leurs mouvements en composant ces cercles de manière différentes, ou de vouloir les expliquer de manières plus compliquées.)

Bref tous ces remèdes sont équivalents et les ellipses peuvent les remplacer . Mais une fois de plus, Guarini ne prend pas position, il constate et termine en disant que l'on peut faire comme l'on veut, cela risque seulement d'être plus long. Il conclut

Corollarium 2

… Et haec sunt omnes appellationes, quas ad motum Planetarum pertinent, quae etiam militant, si loco Deferentis substituas Ellipsim [*Coelestis mathematicae*, Vol. 1, p. 189].

(Corollaire 2

… Et telles sont les appellations pertinentes pour le mouvement des Planètes et qui militent également pour la substitution des ellipses au lieu des Déférents.)

En fin de ce huitième traité, Guarini va introduire l'hypothèse de la Terre mobile.

Expensio VIII : De hypothesi terrae motae.

Quamvis, quod terra moveatur putemus ingenij ostentationem a sensuum ductu se rebellantis, & eorum suasionem dedignanti ; adhuc tamen eam declarare opportet, ne quid huic Tractatui desit et incompertum relinquatur [*Coelestis mathematicae*, Vol. 1, p. 190].

(Expensio VIII : De l'hypothèse de la Terre mobile :

Nous pensons que le mouvement de la Terre est un ingénieux faux semblant contre lequel on se rebelle guidé par les sens; néanmoins il faut ignorer leur persuasion, et déclarer cette hypothèse opportune pour que ce traité ne manque à son devoir et reste incomplet.)

Une telle phrase ne peut s'interpréter que par le refus de son auteur de prendre position. Il va ensuite terminer le traité VIII par une double négation appliquée à un verbe négatif lui-même, à savoir éviter (*eos non vitent necesse non era*) pour laisser le lecteur dans la plus totale incertitude.

Theor. IV. Propos. LIII

… Quare evidens est istos motus terrae Assertores tandem in id consentire, quod summopera effugiunt. Iam enim in Luna Epicicli, seu Ellipsis loco Epicicli cogunt incommodum experiri : unde cum penitus eos non vitent necesse non erat tanto machinamento terra a sua quiete divellere, & in motus extortos cogere, & hae breviter dicta sint de Systematibus terrae

errantis ; Motus enim medios aequationes Apogaeorum motus, &
anomalias, ut supra diximus ordinant, ut quisque ex se poterit facile
cognoscere [*Coelestis mathematicae*, Vol. 1, p. 191].

(Theor. IV. Propos. LIII

C'est pourquoi les défenseurs de ce mouvement de la Terre doivent reconnaître qu'ils manquent leur but. Déjà en effet, ils reconnaissent avoir des difficultés pour la Lune avec les épicycles ou ellipses à la place d'épicycles: donc comme profondément, il ne leur était pas nécessaire de ne pas éviter une telle machinerie pour ôter à la Terre son repos et pour lui attribuer un mouvement tortueux qu'ils appellent rapidement système de la Terre errante ; on ordonne le mouvement au moyen des équations du mouvement des apogées et des anomalies, comme nous l'avons dit plus haut, et n'importe qui peut s'en assurer facilement.)

Tractatus IX : Solis motus decreti

(Traité IX : Dogmes du mouvement du Soleil)

Guarini n'évite donc pas la théorie traditionnelle mais on remarquera le terme *décretum* que nous traduisons pas « dogme », forçant un peu l'aspect imposé. Mais cet aspect est bien présent dans décret également.

Tractatus XV : Duorum inferiorum hypoteses stabilitae

(Traité XV : hypothèses établies pour les deux planètes inférieures)

Theor. II. Propos. II : Duo Planeta inferiores modo sunt supra, modo infra Solem

Probatur ex Phasibus, quae Telescopio detectae sunt. Nam sicut in Luna illuminationes variae sunt, modo plena, modo cornuta, modo gibbosa, appareat, sic & hi Planetae, & praecipue Venus [*Coelestis mathematicae*, Vol. 1, p. 380-381]

(Theor. II. Propos. II : Les deux planètes inférieures sont parfois au-dessus, parfois en dessous du Soleil.

Cela se montre au moyen des phases que le télescope a détectées. Car comme les illuminations de la Lune sont variées, et qu'elle est parfois pleine, parfois en croissant, ou gibbeuse comme ces planètes [inférieures] et principalement Vénus.)

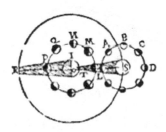

Fig. 12. Dessin rassemblant une Lune tournant autour de la Terre et une planète tournant autour du Soleil pour illustrer le phénomène des phases de la Lune et de Vénus

Dans le quinzième traité, Guarini signale que les planètes inférieures ont comme la lune des phases. Il illustra cette affirmation par un dessin rassemblant une Lune tournant autour de la Terre et une planète tournant autour du Soleil. Il montre ainsi les deux phénomènes de phases et le fait que la planète passe alternativement devant et derrière le Soleil alors que la Lune reste toujours en deçà de ce dernier.

Tractatus XVIII : *Proprietates planetarum expositae*

(Traité XVIII : Exposé des propriétés des planètes)

Expensio I : *De figura planetarum.*

(Expensio I : de la figure des Planètes.)

Guarini rassemble ici les résultats des observations faites au moyen du Télescope, principalement par Galilée, à qui il en attribue erronément la découverte.

Concl. I. Propos. I : *Saturnus ad modum lentis efformatus probabiliter est*

Probatur ex observationibus Galilei, Fontanae anno 1630, Riccioli anno 1643, Nicolae Zucchij, & Ugenij, qui de haec re librum scripsit.

(Concl. I. Propos. I : Il est probable que Saturne soit formée de Lentilles

Cela se prouve grâce aux observations de Galilée, Fontana en 1630, de Riccioli en 1643, de Nicolas Zucchi et de Huygens qui a écrit un livre sur ce sujet.)

Fig. 13. Représentations de Saturne proposées par Guarini, nous y reconnaissons l'étoile triple de Galilée en C les anses proposées par Fontana et Riccioli en A et une forme bizarre en B qui pourrait aussi correspondre à une proposition de Fontana

Guarini résume ici toute l'histoire de la découverte de la forme de Saturne. Initiée par Galilée qui croyait à la présence de trois étoiles accolées, fut recomposée par Riccioli qui rassemble dans son *Novum Almagestum* toutes les formes qui ont été proposées pour Saturne. C'est finalement Huygens qui va expliquer sa forme entourée d'un anneau dans le *De systema saturnium* [1659] (fig. 14).

Fig. 14. Représentation de Saturne par Christiaan Huygens

Concl. IV. Propos. IV : Venus, & Mercurius ad sensum sphaerici sunt.

(Concl. IV. Propos. IV : Venus, & Mercure sont sphériques aux sens.)

Mais le télescope fait découvrir une d'autres formes

Fig. 15 Représentations des phases de Mercure et de Vénus telles que les montre le télescope selon Guarini

Guarini passe ensuite à l'illumination des planètes qui cause le phénomène des phases.

Expensio II de illuminatione Planetarum

(Expensio II de l'illumination des Planètes)

Theor. I. Propos. VII. : *Lunae suum a Sole Lumen mutuatur, & hinc diversis phasibus variatur.*

(Theor. I. Propos. VII. : La lune reçoit sa lumière du Soleil et donc elle a des phases variables.) (fig. 16)

Theor. IV. Propos. X. *: Planetae superiores recipientes lumen a Sole, neque corniculati, nex bifidi cerni queunt ; sed tantum gibbi [Coelestis mathematicae, Vol. 1, p. 444].*

(Theor. IV. Propos. X. : Les Planètes supérieures reçoivent leur lumière du Soleil mais ne sont jamais en croissant, ni en demi, mais parfois gibbeuse.) (fig. 17)

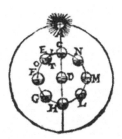

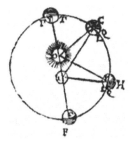

Fig. 16. Les phases de la Lune expliquées par Guarini

Fig. 17. Explication de l'absence de phase des planètes supérieures

Ces deux dernières figures se combinent pour former la Fig. 12 que nous avons vue supra.

Theor. V. Propos. XI. *: Venus, & Mercurius Lumen a Sole recipiunt, & iisdemque phasibus, ac Lunae variantur.*

Probatur vero ex Phasibus, quae tubo optico ut fig. pr. 4 videre licet,

(Theor. V. Propos. XI. : Vénus et Mercure reçoivent leur lumière du Soleil et ont les mêmes phases que la Lune.

Cela est prouvé par les observations fait au moyen de la lunette comme on l'a montré dans la figure supra [cf. Fig. 18].)

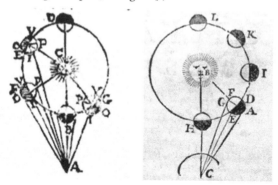

Fig. 18. Les phases de Mercure et Vénus expliquées par Guarini et Fig. 19 par Riccioli

Le phénomène des phases de Vénus est essentiel pour l'héliocentrisme. En effet, les phases de Vénus se succèdent dans l'ordre opposé des phases lunaires comme on peut le voir sur la Fig. 12 de Guarini. Si on ajoute à cela les différences de taille de Vénus au cours de son trajet qui prouvent qu'elle s'éloigne et se rapproche de la Terre au lieu de rester à même distance comme le voudrait la théorie géocentrique, on peut se convaincre de ce que Vénus tourne autour du Soleil. Il est difficile de croire, après les explications précises qui viennent d'être données, que Guarini n'ait pas été sensible à cet argument.

Guarini va terminer son traité XVIII en présentant en parallèle comme ses contemporains, les différents systèmes planétaires et les confronter à ses mesures. Il va au passage proposer son propre système, très proche du système ptoléméen affirme-t-il cette fois haut et fort ; trop haut, trop fort après ce qui précède (voir aussi la Fig. 11 qui reproduit les pages 454-455).

Expensio VI : *De loco planetarum*

(Expensio VI : l'emplacement des Planètes)

Conc. II. Propos. XXVIII :

Planetarum ordo is est supposito terrae centrum occupare ; primo Luna ambit terram, Solque remotior, Venus, & Mercurius eadem Caelesti regione deportantur ; Superiores autem circa terram, & circa Solem in excentrico delabuntur ampliori, quam Solis circuitu [*Coelestis mathematicae*, Vol. 1, p. 454].

(Conc. II. Propos. XXVIII :

L'ordre des Planètes est le suivant, la Terre occupe le centre, la Lune est la première à tourner autour de la Terre, puis vient le Soleil, Vénus, Mercure se trouvent dans la même région du ciel, les planètes supérieures tournent aussi autour de la Terre et autour du Soleil aussi mais de manière excentrique sur des cercles plus grands que celui du Soleil.)

Conc. III. Propos. XXIX :

Caeli Regiones per quas planetae deducuntur latiores sunt, ita quod nec Saturnus gyrando in Orbe, Caelum Iovis occupet, nec Iovis regionem Martis, nec Veneris Lunae : Sed Mars Solis confinia aliquando permeat in perigae orbis, Venus, & Mercurius toto orbe, seu Epiciclito motu semper perambulant [*Coelestis mathematicae*, Vol. 1, p. 454].

...

Praeter hoc nostrum sistema, qui cum Ptolemaico fere convenit, adest Ptolemaicum, quod solum differt a proposito in hoc, quod semper Martem super Solem ferri autumet.

Tychonicum, qui terram orbitae Solis esse centrum statuit, & Lune. Solem vero esse centrum omnium quinque Planetarum [*Coelestis mathematicae*, Vol. 1, p. 454].

...

Sistema Copernicaeum Solem Immobilem in medio Mundi collocat, circa quem terra, omnesque Planetae sub Zodiaco perferuntur [*Coelestis mathematicae*, Vol. 1, p. 455].

(Conc. III. Propos. XXIX :

Les régions du ciel que les Planètes parcourent sont larges, de telle manière que Saturne tournant sur son orbite n'occupe pas le ciel de Jupiter , ni Jupiter celui de Mars, ni Vénus celui de la Lune : Mais Mars est si près du Soleil qu'elle pénètre dans son orbite au périgée, Vénus et Mercure parcourent toujours leurs orbes ou épicycles

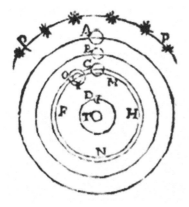

Fig. 20. Système général ptoléméen et système de Guarini

...

... Donc notre système, qui concorde pratiquement avec celui de Ptolémée et qui est Ptoléméen et n'en diffère que par le fait que ce dernier suppose que Mars est toujours au-dessus du Soleil.

...

Le système de Tycho qui affirme que la Terre est le centre de l'orbite solaire et de celle de la Lune. Mais que le Soleil est le centre de l'orbite des cinq autres Planètes.

... Le système de Copernic place le Soleil immobile au centre du Monde, autour de lui se déplace la Terre et toutes les autres Planètes sur le Zodiac.)

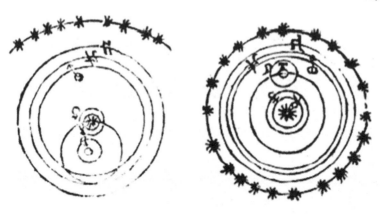

Fig. 21. Le système de Tycho Brahe   Fig. 22. Le système de Copernic

Conclusion

Je ne tenterai pas de conclure en affirmant une quelconque prise de position par Guarini à propos de l'héliocentrisme, alors que je pense avoir montré qu'il refuse de prendre une telle position. Je voudrais seulement montrer le pavement de la Capella della Sindone parce qu'il est tellement beau et parce qu'il contient, on peut y rêver la solution de cette énigme (Fig. 23).

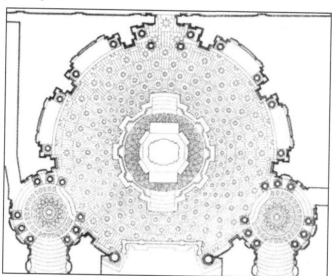

Fig. 23. Pavement de la Capella della Sindone

Bibliographie

GUARINI, Guarino. 1665. *Placita philosophica*. Parisiis, apud Dionysium Thierry.

————. 1666. *Trattato di fortificatione, che ora si usa in Fiandra, Francia, & Italia*. Torino, Per gl'Eredi Gianelli.

————. 1671. *Euclides adauctus et methodicus mathematicaque universalis*. Turin, Zapata, 1671 réédit 1676.

————. 1674. *Modo di misurare le fabbriche*. Torino, per gl'heredi Gianelli.

————. 1675. *Compendio della sfera celeste*. Torino, appresso Giorgio Colonna.

————. 1678. *Leges temporum et planetarum*. Torino, Per gl'Eredi Gianelli.

————. 1683. *Coelestis mathematicae pars prima et secunda*, Mediolani, ex Typographia Ludouici Montiæ.

————. 1686. *Disegni di architettura civile ed ecclesiastica*. Torino, Per gl'Eredi Gianelli.

————. 1726. *Novum Theatrum Pedemontii et Sabaudiae...*, Hagae-Comitum: sumptibus et cura R. C. Alberts.

————. 1737. *Architettura civile*. Torino, G. Mairesse all'insegna di Santa Teresa di Gesú.

HUYGENS, Christiaan. 1659. *Systema saturnium, sive de Causis mirandorum Saturni phaenomenón et comite ejus planeta novo*, Hagae-Comitis : ex typ. A. Vlacq.

L'auteur

Patricia Radelet-de Grave is a physicist. She teaches history of mathematics and physics to mathematicians, physicists, engineers, architects and computer scientists at the Catholic University of Louvain la Neuve. She is also the general editor of the Bernoulli Edition. Her main interests are the relations between mathematics and physics, which, for example, she studied through the interactions between the concept of vector and that of force, or between the mathematical concept of field and the concept of electromagnetic field. Studying forces, she was led to works on stability of vaults. She then met Edoardo Benvenuto, with whom she initiated the research series *Between mechanics and architecture*.

Clara Silvia Roero

Dipartimento di Matematica
"G. Peano"
Università di Torino
Via Carlo Alberto, 10
10123 Torino ITALY
clarasilvia.roero@unito.it

Keywords: history of
mathematics, Guarino
Guarini, Pierre Hérigone,
Gaspar Schott, Claude-
François Milliet Dechales,
Christopher Clavius,
Bonaventura Cavalieri,
Grégoire Saint Vincent

Research

Guarino Guarini and Universal Mathematics

Abstract. Guarini considered the mathematical studies to be of fundamental importance for all artists and scholars. His own knowledge of mathematics was vast and profound. The aim of this present paper is to show, through an analysis of the most substantial of his mathematical works, *Euclides adauctus*, along with the *Appendix* to this printed a few months later, the role that philosophical and mathematical studies had on his cultural formation, on the new and original research that he conducted, and on his teaching activities, while looking for traces of the mathematical sources that he consulted and cited that indicate which authors and works exerted the greatest influence on him.

Introduction

Guarino Guarini, in the dedication to Charles Immanuel II of Savoy of his most important scientific treatise, *Euclides adauctus et methodicus mathematicaque universalis*, printed in Torino in 1671, underlines the miraculous power that mathematics exerts on architecture, declaring that it is possible to draw on mathematics' most sublime ideas, a science that he sought to enrich with the fruits of his labour:

> ...but above all it is architecture that shines thanks to the distinguished and truly regal Thaumaturgy of the miracles of mathematics. ... Hereby receive, your Royal Highness, with benign visage and serene clemency, that which several times with the breadth of its ingeniousness in conceiving the most sublime ideas has fostered mathematics and all of the efforts of my work in adorning it.[1]

Guarini's predilection for mathematics over all other disciplines of learning is expressed more than once. It manifests itself concretely in his artistic creations, which make visible the beauty and harmony of forms born from his love of plane and solid geometry (cf. [Roero 2001]).

In his first work of a philosophical-scientific nature, *Placita Philosophica* (fig. 1), published in Paris in 1665, he upholds the importance of the knowledge of mathematics for all artists and scholars:

> All the arts depend on either mathematics, philosophy or medicine, all sciences that examine similitude, proportion or the fittingness of things ... Thus, the more profound the artist's knowledge is regarding the things relative to his art, of the means and manners of applying them, the more excellent will he be judged, and the more perfect his works will be considered. In fact, when the artist sets himself to his task, he does well to choose the most suitable material, to have a perfect knowledge of his instruments, and the ties with all the things relative to the art, and finally to eliminate the devices used to create each thing. And since in the most difficult situations neither imagination nor intellect is sufficient, the devices drawn from these models can be applied most perfectly to the idea to be demonstrated and performed in that particular circumstance.[2]

Nexus Network Journal 11 (2009) 415–439   NEXUS NETWORK JOURNAL – VOL.11, No. 3, 2009 **415**
DOI 10.1007/s00004-009-0012-x; *published online* 7 November 2009
© 2009 Kim Williams Books, Turin

Fig. 1. Frontispiece of Guarini's *Placita Philosophica*, 1665

In his final work, *Architettura Civile*, published posthumously in 1737 under the direction of Bernardo Vittone, Guarini explicitly states architecture's dependence on mathematics:

> Of the, we might say, infinite operations that mathematician perform with explicit demonstrations, we will select some of those of prime importance that are necessary to Architecture, without however providing the proofs, because this is proper to Mathematics, of which Architecture professes to be a disciple.[3]

In order to arrive at a better understanding of the elements that inspired Guarini's architectural constructions, it seems opportune to look at the extent and depth of his knowledge of mathematics and how this influenced his artistic activities. In spite of the fact that Guarini has been studied by historians in various fields of culture, and that these have shed light on the legacy he received from other architects both earlier and contemporary, until now few have undertaken an analogous study in terms of his mathematical learning and the relationships between his knowledge of geometry and his works, both written and built. The aim of this present paper is to show, through an analysis of the most substantial of his mathematical works, *Euclides adauctus*, along with the *Appendix* to this printed a few months later, the role that philosophical and mathematical studies had on his cultural formation, on the new and original research that he conducted, and on his teaching activities, looking for traces of the mathematical sources that he consulted and cited in order to identify the authors and works which exerted the greatest influence on him. We will illustrate the structure, characteristics and aim of this treatise, note the elements that are new, and its limits with respects to similar contemporary works. Finally, we will consider the diffusion of the work, and its legacy in scientific and pedagogical terms.

Guarini and the Euclides adauctus

A cosmopolitan artist, born in Modena on 17 January 1624, after his formative years in the colleges of the Theatines in Modena and Rome, where he studied theology, philosophy, mathematics and architecture, Guarini had the good fortune to come into contact with various societies and cultures during his travels to Rome, Vicenza, Prague, Parma, Guastalla, Messina, Lisbon, Paris, Nice, Torino and Milan. His cultural baggage was immensely enriched by encounters with scholars, artists, teachers, librarians and rulers in various countries. The fruit of these exchanges ripened during Guarini's period in Torino, from 1666 until his death on 6 March 1683. These years were undoubtedly the most fertile and rich in initiatives, during which he realised, in addition to admirable works of architecture, no fewer than six printed works [Guarini 1671, 1674, 1675, 1676, 1678, 1683] on topics of mathematics, stereotomy, geodesy, gnomonics (the study of sundials) and astronomy, as well as two works on architecture, which were published posthumously in 1686 and 1737. The majority of these publications reflect Guarini's remarkable mathematical gifts, recognised early on by Charles Immanuel II of Savoy, who in May 1668 named him Royal Engineer and Mathematician.

Fig. 2. Frontispiece of Guarini's *Euclidis Adauctus,* 1671

Euclides adauctus (fig. 2) is on a much higher level than other treatises of geometry and practical arithmetic assembled in the Renaissance for artists and artisans by authors of note such as Leon Battista Alberti, Francesco di Giorgio Martini, Leonardo da Vinci and Albrecht Dürer, although in a certain sense it shares with them, as we shall see, a similar purpose.[4] Despite its title, it also differs clearly from the various editions of Euclid's *Elements*, whose aims were of a philological and critical nature, and which were quite widespread in Italy in the sixteenth century.

Guarini's work is encyclopaedic in nature and is in some way comparable to Luca Pacioli's *Summa de arithmetica, geometria, proportioni et proportionalitate* [1494] and to Niccolò Tartaglia's *General Trattato di numeri e misure* (4 vols., 1556-1560), especially with regards to the readership it was aimed at. As Guarini himself stated explicitly several times – in the *Euclides*, in the *Modo di misurar le fabriche* and in *Architettura civile* – his aim was to make known the results achieved in classical geometry as found in the works of Euclid, Archimedes, Apollonius and Pappus, and of more recent mathematics regarding, for example, curves, solids of rotation, indivisibles and logarithms. His works were destined for a readership that was learned and exigent, but not one specialised in subtle and complex mathematical proofs. His public comprised intellectuals and practitioners desirous of understanding the fundamentals of geometry and arithmetic in order to apply them, for example, in fields such as architecture, geodesy, military architecture, gnomonics and astronomy. It was for this reason that in the introduction to chapters addressing a specific topic, he often made mention of possible applications for the particular theory [Guarini 1671: 26]. For example, he mentions the use of regular polygons to construct fortifications [Guarini 1671: 83]; he says that the conics can be used to fabricate burning glasses and sundials, and to study the motions of the planets [Guarini 1671: 422]; he notes that projections are used in gnomonics and architecture, in building sundials and mathematical instruments such as the astrolabe and the quadrant, and in cosmography [Guarini 1671: 444]. Further, Guarini conceived new curves, new surfaces and new solids which could be used in the construction of buildings, churches, noble palaces, gardens, staircases, columns, vaults, lunettes, and more. He also mentions applications of mathematics to geodesy [Guarini 1671: 503].

From the very beginning of the *Euclides adauctus*, indeed on the frontispiece, Guarini states the works nature and contents:

> Euclid amplified and set out methodically, and universal mathematics dedicated to the Duke of Savoy, Prince of Piedmont, King of Cyprus, etc., Charles Immanuel II, which not only observes the dependence of the propositions, but also the order of things. This work is even complete with all of those properties that can be observed with regards to both discrete quantities and abstract continuous ones. Superfluous demonstrations have been neglected and all the important ones included; further, the individual treatises have been amplified with new propositions and some parts have been entirely rewritten. All is illustrated clearly and precisely with figures and with words.[5]

In the preface to his 'gentle readers', Guarini uses a beautiful metaphorical image of the reasons why he was led to compile the treatise, that is, that his own personal experience had confirmed "the value and usefulness that this kind of work can have to *irradiate with mathematical light and make evident all things with a single luminous source*".[6]

The inspiration for this came to him from the encyclopaedic collections called *Cursus Mathematicus* published in France and Belgium, such as that of Pierre Hérigone (1580-1643), in five volumes, printed in Paris [1634-1637],[7] and that of Gaspar Schott (1608-1666), in twenty-eight books, published in Würzburg (in Latin, Herbipoli Franconiae) [1661].[8] Guarini most likely came to learn of Hérigone's work during his travels in France, in Paris, where Hérigone was much appreciated as a teacher. Guarini may have had direct contact with Schott, a Jesuit, in Sicily, because Schott taught at the Gymnasium of the Jesuit College in Palermo. Schott later transferred to Würzburg, where he published, in addition to his *Cursus Mathematicus sive absoluta omnium mathematicarum disciplinarum encyclopaedia*, the volumes entitled *Mathesis Caesarea* [1662a] and *Physica curiosa sive mirabilia naturae et artis* [1662b], which shows his delight in writing treatises of a practical-applicative nature aimed at a curious public whose vast interests included the "mathematical magic and instruments" of Athanasius Kirker; compasses and other equipment used for measuring in geodesy and polygraphy, that is, in geometry applied to military architecture; the firing of projectiles; problems of military tactics; optics; currency exchanges; chronology and the calculation of the date of Easter; mechanics; meteorology; astronomy and civil architecture.

In the first edition of *Euclides adauctus*, in 1671, Guarini did not yet know of the celebrated *Cursus seu Mundus Mathematicus* by the French Jesuit Claude-François Milliet Dechales (1621-1678), which was published in three volumes in 1674. But this important work was cited by Guarini in *Architettura civile* [Guarini 1737: 1].[9] In all likelihood Guarini and Dechales met in Torino in 1674, when Dechales gave a scholarly lecture in the *aula magna* of the Jesuit College, which four years later was also the site of a solemn memorial service for Dechales, with a funerary oration that was published in Torino.[10]

Guarini proudly underlines that he had avoided the disadvantage of spreading the notions over many costly volumes, and that he had collected the concepts and properties in an orderly way and in succession in a single work, especially for "those who are not able to translate the most difficult knowledge into their own language". He mentions the difficulty that the ordinary reader, that is, one who is not a professional mathematician, would face if they were to take on the works of Apollonius of Perga, Archimedes and Pappus directly. In his treatment, although he had consulted many texts, he did away with what was superfluous, which he judged to be useless, and did not include all the discussions of the ancients; he sometimes broadened the scope of that mathematics, by inserting proofs that were missing; he often improved on proofs, and he always marked the innovations and changes that were his own with an asterisk in the left margin, since these were the fruits of his personal reckoning and reflection. In this volume Guarini reveals his excellent gifts as a teacher, stating that he himself had learned the way to explain mathematics from Euclid and Proclus:

> Thus the ones who are most guilty of violating the laws of mathematics, as we learn from the Lion's claw [i.e., Proclus] are those who, while proving a proposition, insert, or even only state, other observations and propositions that are out of place there, and that have not already been proven, and in so doing they confuse the students' minds and cause them problems.[11]

The style of the work in fact shows a remarkable sensitivity for didactics, originality and depth in the explanations, and singular skills in expression, with great attention paid to how the properties can be visualised, which derives from Guarini's extraordinary experience in artistic fields.

The topics addressed are subdivided into thirty-five books or treatises (*tractatus*), which are in their turn subdivided into chapters (*expensio*), and which include definitions, postulates, principles, theorems, corollaries, problems and, occasionally, "assumptions" (*praeassumptum*) and conclusions (*conclusio*). Each treatise opens with a foreword in which Guarini mentions that particular topic's possible practical uses by engineers, artisans, instrument makers, military architects, geodesists, musicians, and so forth. The way the subject is set out is closer to that of Euclid's *Elements* than it is to the works of Archimedes and Apollonius, which were addressed to readers who were highly cultured experts. The proofs are clear and detailed. Guarini numbers in order, sometimes excessively so, the steps to be taken.

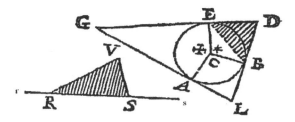

Fig. 3. Guarini 1671, Tract. VII, Probl. II Prop. III, p. 85

He also makes use of efficacious graphic aids in order to better visualise the elements under consideration. These include the insertion of asterisks or crosses in angles (fig. 3); the hatching of angles, faces, planes or surfaces, which are then referred to in the text as "black angle" or "black rectangle" (fig. 4), etc.; the intersection of planes to which a thickness is given (fig. 5), the drawing of parallel circles or rings on spherical surfaces (fig. 6); the edges of regular polyhedra (fig. 7), imitating the *poliedri vacui* drawn by Leonardo da Vinci for Luca Pacioli's *De divina proportione*, prospective projections. The illustrations are perfectly drawn down to the smallest detail.

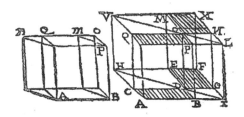

Fig. 4. Guarini 1671, Tract. XXXIV, Th. V, Prop. VI, p. 611

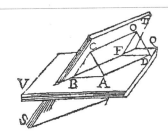

Fig. 5. Guarini 1671, Tract. XXII, Def. IV, p. 347

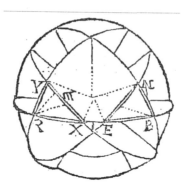

Fig. 6. Guarini 1671, Tract. XXIII, Exp. IV, Th. I, Prop. XI, p. 363

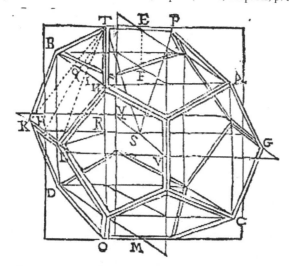

Fig. 7. Guarini 1671, Tract. XXXII, Exp. VI, Probl. I, Prop. IX, p. 603

The first three treatises of the *Euclides adauctus* reintroduce arguments of a philosophical nature already addressed in the *Placita Philosophica* regarding the existence of continuous quantities, discrete quantities, indivisibles, the infinite, and the characteristics of mathematics.[12] Before taking on the mathematical theory of proportions, Guarini, following Aristotle's theories in *Physica* and *Metaphysica*, studies actual and potential, illustrating their metaphysical properties and distinguishing between six kinds or species of quantity: mass, number, time, motion, virtue and weight. In this work Guarini only takes into consideration quantities that are continuous – the object of geometry, points, lines, surfaces – and discrete – the object of arithmetic, numbers.

Here Guarini spends a great deal of time discussing the existence of divisible or indivisible points in quantities and in the theory of atomism. He addresses the form of atomism that maintains that the various bodies are composed of indivisible atoms that are separate and contiguous, such that between any two there are no other intervening atoms. He refutes this kind of concept by citing some considerations already made in the *Tractatus de continuo* by Thomas Bradwardine of Merton College in Oxford regarding concentric circumferences and the side and diagonal of a square, arguing that if biunique correspondence between the points of these entities were established, then in the case of concentric circumferences, the lengths would be equal, while in the second case the magnitudes would not be incommensurable, since the side and diagonal of the square would have the same number of indivisible atoms.

The various arguments reflect Guarini's knowledge of medieval and Renaissance debates regarding the continuous, the polemic between Jacques Pelletier and Christopher Clavius concerning the angle of contact between circumference and tangent,[13] the use of indivisibles in mathematics reasoning, with examples drawn from Galileo's *Dialogo sopra i due massimi sistemi del mondo*, Luca Valerio's volume on the hemisphere *De centro gravitatis solidorum* [1661], Vincenzo Viviani's *De maximis et minimis* [1659] and Bonaventura Cavalieri's *Geometria indivisibilibus continuorum* [1645].

The treatment of the infinite is quite detailed from the point of view of both mathematics and physics. Like Aristotle, Guarini refutes actual infinity, maintaining that it is not possible to identify the last part of such an infinity, because it is an infinity that cannot be exhausted.[14] By means of well-chosen examples from the best literature of the day, Guarini shows why actual infinity cannot be admitted. In addition to Galileo's reflection on the paradoxes of infinity, Guarini also includes those of St. Augustine on the fact that it is not possible to assign an infinite number, and that only God knows infinity, can see all infinite points, and is capable of selecting and separating them. For Guarini it is possible to conceive parts of quantities that exist in infinity, as long as such quantities are conceived of "mathematically", that is, in the abstract. Incommensurable magnitudes are divisible at infinity, and there is no common unit, thus there is no minimum magnitude; this proves the existence of potential infinitesimal magnitudes. That is to say, there are infinities and infinitesimals that can be conceived in the mind. For example, a sheaf of parallel lines intersected by transversals lead to obtaining various degrees of magnitude of increasingly smaller dimension, to infinity; the angle of contact can be divided infinitely; in the quadratrix of Dinostratus it is possible to successively divide the angle, but the final point of the curve will never be obtained. Guarini then asks if the physical point is divisible infinitely, and provides an answer of a philosophical and theological nature. Finally, he indicates what the mathematical indivisibles are: point, line, surface. He gives a particularly interesting definition of a point: *cujus pars nulla* ("that whose part is zero"), a notion that would be picked up by Deshales in his *Cursus seu Mundus Mathematicus*. The treatise concludes with the chapter discussing "if indivisibles can be the object of mathematics", in which Guarini comments on the work of Cavalieri (1598-1647), praising his intellect and his doctrine of indivisibles:

> Bonaventura Cavalieri dedicated himself to furthering mathematics by examining indivisibles with intelligence and acumen in a book dedicated to this purpose and to finding equalities and proportions in regard to indivisible points that exist in quantities.[15]

Here he cites both the objections to the method of indivisibles used by Mario Bettini in the *Epilogus Planimetricus*[16] and that of Paul Guldin in *De centro gravitatis solidorum* [1642: **-**], as well as the authors who appreciated the mathematical proofs, such as Ismaël Bullialdus in *De lineis spiralibus*[17] and Vincenzo Viviani in *De maximis et minimis* [Viviani 1659: Lib. I, Theor. IX, Prop. XVII, Monitum, 35]. Guarini's conclusion is articulated in nine points and ends with the judgment that Cavalieri did not provide an actual and evident proof because in his method he goes from one species to another: the indivisible segments (of the first species) form a surface (of the second species) and this kind of proportion between figures of different species is not permitted in geometry.

Considerations of a philosophical nature, drawn from Aristotle and from his medieval and contemporary commentators, are also present in the second treatise regarding the nature of discrete quantities, the number one, and whether or not an infinite number exists [Guarini 1671: 13-20]. Here Guarini displays a profound knowledge of Aristotle's *Metaphysica* and contests some of the statements of the Persian Avicenna (Ibn Sina, ?-1037) and the Jesuits Pedro De Fonseca (?-1599), Francisco Suarez (?-1617) and José Maria Suarez (?-1677), whom he may have met during his travels in Spain.

In treatise III Guarini addresses the properties of the discipline of mathematics, and gives special attention to its philological aspect and to the the terminology used by the ancients (Pappus, Proclus) and the moderns (Clavius, Pierre de la Ramée, Girolamo Vitali) to indicate the elements, definitions, principles, postulates, theorems, problems, lemmas, and so forth. He also discusses the importance of teaching mathematics in education.[18]

After having noted that "the name Mathematics derives from the Greek and means doctrine or discipline in Latin", he explains that,

> ...it teaches only by means of demonstration (*per ostensionem*) and declares to be proven only what is evident and deduced from the principles. Mathematics is a *ostensive science* whose object is all that is measurable [Guarini 1671: 22].

Guarini distinguishes between three types of mathematics – universal, cosmic and microcosmic – and declares his intention in this work to deal with only the first of these, universal mathematics, which is its turn subdivided into arithmetic and geometry, since these open the doors to all the others.[19] In treatises IV-XII Guarini presents and proves the propositions set forth by Euclid in books I-VII and X of the *Elements*. In treatises XXII and XXXIII he considers solid geometry, the intersection of planes and the inscription of the five regular polyhedra in the sphere, a theme addressed by Euclid in his books XI, XII and XIII.

For book V of the *Elements*, dedicated to the theory of proportions, Guarini reserves his treatises VIII and IX, in which he addresses the arithmetic operations and the proportions of segments. The particular types of proportions that Euclid considered in his books VI and VII are deal with by Guarini in his treatises X and XI. In his treatise XII he then addresses the incommensurable magnitudes and irrational numbers, which Euclid set out in his book X. Instead, the operations involving fractions and the rules commonly used to solve some problems of arithmetic, such as the golden rule, the simple rule of three, the composite rule of three, and so forth, and the algorithm for extracting the square root and cube root are examined in treatise XIII. Finally, particular kinds of means and progressions – the arithmetic, geometric, harmonic and so forth – are addressed in treatises XIV, XV, XVI and XVII.

The sources that Guarini consulted in compiling all of these treatises were the classic editions of Euclid's *Elements* in Latin edited by Christopher Clavius[20] and Francesco Commandino,[21] and those in Italian by Commandino [1575] and by Niccolò Tartaglia,[22] but also the texts on spherical trigonometry by Theodosius, Ptolemy, Menelaus, Copernicus and Regiomontanus (cf. [Guarini 1671: 347, 360]). However, he also read and cited authors of manuals of arithmetic and practical geometry, such as

Giovanni Antonio Magini,[23] Mario Bettini,[24] again Clavius[25] and Giambattista Benedetti (1530-1590) on the elementary geometrical constructions with a compass with a fixed opening.[26]

In treatise VII particular attention is given to constructions with straightedge and compass, for example, constructing the sum and difference of segments, the bisector of an angle, the perpendicular or parallel to a given line and the reciprocal inscriptions and circumscriptions of regular polygons in the circle.[27] Guarini displays a knowing mastery of the subject, derived from his studies of Clavius's version of Euclid, from which he adopted numerous corollaries and observations. He sometimes adds his own personal considerations, stating for example in *Expensio V* that Euclid neglected the construction of the hexagon circumscribing the circle and that of the circle inscribed in and circumscribing the hexagon. Further, Guarini notes that some polygons, such as the heptagon, are impossible to construct with straightedge and compass, and underlines the fact that there is no treatment in Euclid's *Elements* of measurements of angles, which Guarini considers a useful complement to the constructions, for which reason he includes it in his treatises XIX and XX. To the usual considerations of the characteristic properties of inscriptions and circumscriptions of the circle with the equilateral triangle, square, pentagon, hexagon and a polygon with fifteen sides, Guarini always adds the proof that the suggested construction leads to a polygon in which all sides and all angles are equal. He is particularly attentive to the rigour of the proofs and is never settles for intuitive explanations. At the end of the chapter he mentions the relations that there are between the vertices of the polygons inscribed by 3, 5 and 15 sides (fig. 8) and by 4, 6 and 24 sides, and suggests a rule for constructing polygons generated by the multiplication of the sides of preceding polygons, drawn from Clavius (cf. [Clavius 1574: III, 26: 122] and [Guarini 1671: 91]).

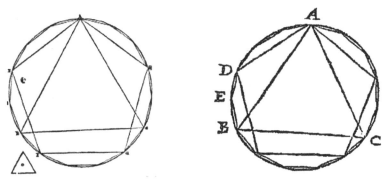

Fig. 8. Polygons of 3, 5 and 15 sides inscribed in a circle in Clavius and Guarini

Guarini sometimes cites specific sources, as in the case of logarithms, set forth in treatise XXI, where he credits John Napier and Henri Briggs as the inventors of that "marvellous and most useful invention" [Guarini 1671: 324].

Of particular interest because of their relationship ties to his built work, are the constructions of the mean proportional between two given segments [Guarini 1671: XV, *De linearum, segmentorumque proportionibus*, 248-255]; the studies on curves and their constructions given in treatise XVIII [Guarini 1671: *De flexis*, 286-299]; and those regarding the conic sections of Apollonius presented in treatise XXIV [Guarini 1671: 390-435]. In numerous points the influence of the *Geometria pratica* by Clavius is evident for the constructions of the proportioning curve [Clavius 1612: vol. 2, Lib. 6, prop. 15, 160-161; Guarini 1671: 249], the quadratrix of Dinostratus[28], and the oval.[29] On the other hand, relative to the construction of the ellipse (fig. 9) Guarini goes back to the works on gnomonics, published by Clavius in Rome,[30] François d'Aguillon in Belgium [D'Aguillon 1613: 465-475], Claude Mydorge in France [1639: 201] and Benedetti in Torino (fig. 10).[31]

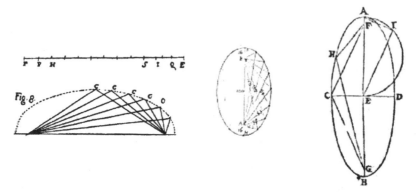

Fig. 9. Constructions of ellipses in Clavius and Guarini

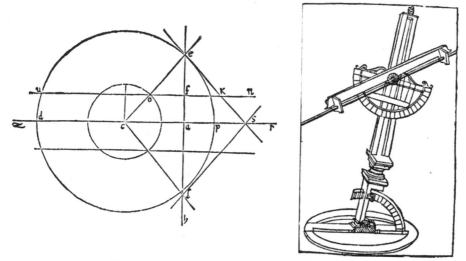

Fig. 10. Benedetti, *De gnomonum ...* 1574, p. 117v

Chapter VII, *De linea conchili*, deals with the conchoid of Nicomedes, which Guarini indicates with the terminology used by Clavius in *Geometria Practica* [Clavius 1612: vol. 2, 162-163]. After noting that this curve was used in the solution of the problem of trisecting the angle (cf. [Guarini 1671: 302-303]), Guarini says that it can be used in architecture for determining the entasis of a column, as suggested by Jacopo Barozzi da Vignola [Guarini 1671, p. 298]. Finally, in chapter VIII, *De lineae ciclicae descriptione*, he mentions the construction of the cyclical line, that is, the cycloid, but does not, however, cite any of the well-known mathematicians, his contemporaries, who studied it, such as Gilles Personne de Roberval, Evangelista Torricelli, Christiaan Huygens and Blaise Pascal.

The treatment of the conics, which are dealt with in treatises XXIV and XXV, is thorough and deep, and follows the classic theory of Apollonius, with the definitions, theorems, properties of tangents, asymptotes, and the famous results of Archimedes regarding the parabola and solids of rotation, and those of Gregorius Saint Vincent in *Opus geometricum quadraturae circuli et sectionum coni*, published in Antwerp in 1647. Oddly enough, Saint Vincent is always referred to as Ambrosius a S. Vincentio,[32] perhaps because Guarini did not have Saint Vincent's work at his disposal, but only those of his student, d'Aguillon.

In the area of solid geometry, Guarini studies the intersections of plane surfaces with bodies generated by the rotation of triangles (obtaining the classic right circular cone), ellipses, parabolas, hyperboloids, and hyperbolic conoids [Guarini 1671: XXV, *De sectionibus sphaericorum*, 436-443]. Guarini also considers a particular coniform solid having a circular genetrix, which instead of terminating in a vertex, terminates in a line segment [Guarini 1671: 438-439]. The practical uses of this kind of geometry are found in sundials, instruments for cosmography, astronomy and above all in architecture, where it is necessary to construct projections. Guarini dedicates treatise XXVI to this theme, which is subdivided into two parts, *De orthographia* and *De stereographia* [Guarini 1671: 444-452, 452-462]. After having addressed the properties of plane and spherical trigonometry in treatise XXVII [Guarini 1671: 463-494], Guarini examines the geometric series relative to surfaces [Guarini 1671: XXVIII, 495-502], the geometric problems of geodesy and the properties of isoperimetric figures [Guarini 1671: 503-526].

The determination of the areas of figures with curved perimeters, compared with others whose perimeters are straight lines, is the object of treatise XXX, in which Guarini mentions the history of the problem of squaring the circle and the results proposed by the ancient philosophers (Antiphon, Bryson of Heraclea, Hippocrates of Chios and Archimedes), as well as by the moderns (Oronce Finé, Nicolaus Cusanus, Saint Vincent, Leotaud and Xavier Franciscus Ayscon) [Guarini 1671: 527-549]. To determine the area of an unknown figure he illustrates the classic method of exhaustion, which consists in the inscription and circumscription of rectangles in the figure such that the difference between the sum of the circumscribed rectangles and the area of the figure is less than any given quantity. Guarini then proves Archimedes' result regarding the measurement of the circle and the approximation for π (that is, the relation between the circumference and the diameter). He then obtains the surface of a circular ring, the area of a lune, the triangle of maximum area inscribed in the ellipse, the area of a segment of parabola found by Archimedes, the area of the Archimedean spiral at its first revolution, an approximation of the area of a segment of hyperboloid. In treatises XXXI and XXXII Guarini deals with the surfaces and volumes of prisms, cylinders, circular groins, cones, truncated cones, elliptical spheroids, spheres [Guarini 1671: 550-571] and their projection on the plane [Guarini 1671: 572-596].

The inscription and circumscription of regular polyhedra in the sphere are addressed in treatise XXXIII [Guarini 1671: 597-608], where Guarini does not settle for merely setting out Euclid's theory, but also extends it in certain points. For example, while in proposition XI.26 Euclid examines the construction of a solid angle equal to a given angle, taking into consideration only the case of a solid angle with three vertices, Guarini states that it is possible to extend the problem to angles of more than three vertices, since these can always be decomposed in solid angles of three vertices [Guarini 1671: 598-599]. After having shown the relation between the sphere and the sides and diameters of each regular polyhedron, Guarini offers a simple and immediate proof of the uniqueness of the five regular solids, based on the initial proposition that in order to construct a solid angle the plane angles that meet at its vertex must necessarily be less than four right angles. He then proceeds to the determination of the volumes of these solids and to the relation that they have with that of the sphere, derived from the work of Archimedes.

The most innovative results are contained in treatises XXXIV and XXXV regarding the surfaces and volumes of bodies not addressed by other mathematicians, and in the twelve-page *Appendix*, added to the work shortly after 1671, described thus:

> Appendix to the Euclides adauctus by Guarino Guarini clerk regular of the Theatines. Because following the printing of the book I was able to arrive at many results regarding the determination of the volumes of solid bodies which no one had yet discovered and examined and these were not only useful but almost required, and above all in practical stereotomy there were no square shells of any kind, nor vaults comprising several bodies, which I had inserted in this work in the part about solids, I added these to those in order that there would be no body bounded within a given surface that was not subject to the measure of solid bodies nor that whose measurement was not determined exactly with mathematical certitude.[33]

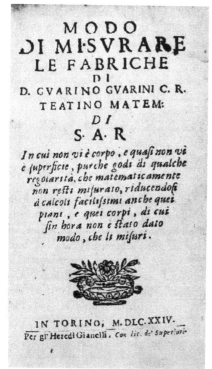

Guarini often refers to this appendix in his next work, the book *Modo di misurar le fabriche* [1674] (fig. 11), where in the third chapter of the second part he proudly reports the theorems of surfaces and volumes of solids that are particularly useful in architecture, for example, "on cloister vaults and lunettes", writing:

> The rules that we will give in this and the chapter that follows regarding the measurement of Vaults based on squares or other figures, with the exception of circles, are all of my own invention, and for which no rules have yet been given [Guarini 1674, p. 101].

Fig. 11. Frontispiece of Guarini's *Modo di misurar le fabriche*, 1674

In the final two treatises of the *Euclides adauctus* and in the *Appendix* Guarini deals with the volumes of bodies contained by plane surfaces, such as pyramids and prisms, and by curved sufaces. Among the curved surfaces he distinguishes three different types: those enclosed by curved sufaces but whose bases are planes with linear boundaries, such as the square fornix; those whose bases are planes with curved boundaries, such as the cone, cylinders and the so-called rhomboid-solid (fig. 12);[34] and those that originate from a surface which is completely curved, such as the sphere, parabolic conoid (fig. 13), hyperboloid, spheroid, and solid rings. In their turn, solid rings can have a base that is square or polygonal; this holds true for the conoids as well.

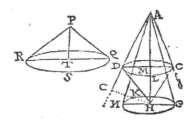

Fig. 12. Guarini 1671, Tract. XXXIV, Exp. II, Th. II, Prop. XV, p. 627

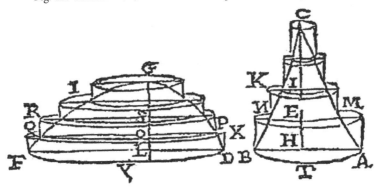

Fig. 13. Guarini 1671, Tract. XXXIV, Exp. IV, Th. V, Prop. XXIX, p. 633

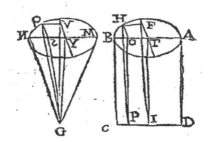

Fig. 14 Guarini 1671, Tract. XXXIV, Exp. IV, Th. I, Prop. XXVI, p. 631

In addition to the right and oblique circular cones, Guarini also examines cones that terminate in a line segment (fig. 14), cones with elliptical bases and coniform solids (fig. 15). The volumes of all these solids are deduced by applying the method of exhaustion illustrated at the end of the preceding treatise (fig. 16).

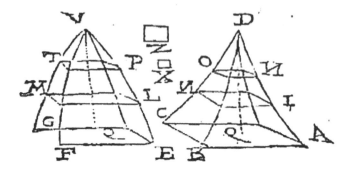

Fig. 15. Guarini 1671, Tract. XXXIV, Exp. IV, Th. III, Prop. XXVIII, p. 632

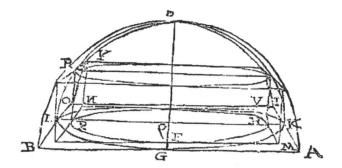

Fig. 16. Guarini 1671, Tract. XXXIV, Exp. IV, Th. III, Prop. XXVIII, p. 632, Guarini 1671, p. 647-648

Guarini then considers surfaces and volumes of semi-spheres with square bases (fig. 16) and conoids or semi-spheroids with square bases (fig. 17), the perimeters of these solids with respect to a circumscribed cylinder (fig. 18), the solids derived from other solids, such as the lunette (fig. 19), the half-quadriforms, spiral-form bodies (fig. 20), and ther relationships among them (treatise XXXV and Appendix).

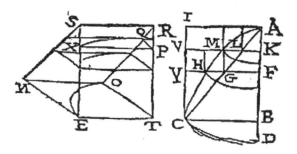

Fig. 17. Guarini 1671, Tract. XXXIV, Th. VII, Prop. LV, p. 652

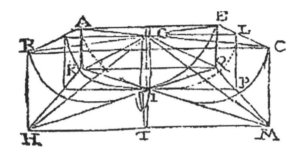

Fig. 18. Guarini 1671, Tract. XXXIV, Th. IV, Prop. LI, p. 649

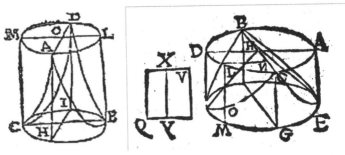

Fig. 19. Guarini 1671, Tract. XXXIV, Exp.XI, Probl. I, Prop. LII, p. 650; Guarini 1674, p. 38

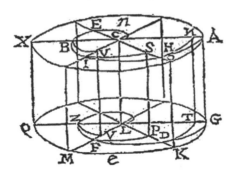

Fig. 20 Guarini 1671, Exp. IX *De corporibus spiralibus*, p. 654; Guarini 1674, p. 191

Conclusion

Given the wealth of new results rigorously deduced by Guarini in these final chapters of the *Euclides adauctus* and in the *Appendix*, in keeping with the style of Euclid, Archimedes, Grégoire Saint Vincent, Luca Valerio, Cavalieri and Viviani, it is striking to note Guarini's mathematical skills applied to figures pertinent to architecture and art. It is precisely in this melding of passion for mathematics and sensitivity to mathematics teaching that the originality of this great mathematical artist is found. From sources both ancient and contemporary he was adept at selecting, collecting and systematizing the principle results, proving them in a rigorous fashion, somes adding his own personal observations relative to the generalization of the properties and the theories set out by other authors. Even though his intended readership was not composed of mathematicians, and comprised students at religious colleges or universities, engineers and architects, draftsmen and artisans, and instrument makers, Guarini did not limit himself to stating theorems or problems, or to describing properties in a superficial way, as did some of his contemporaries (for example, [Magini 1592; Bruni 1967; Bettini 1647-48; Caramuel 1670]) but rather, skillful teacher that he was, he chose to satisfy the demands of the most capable, the most impassioned and the most curious.

Neither Dechales nor Tricomi were capable of identifying the innovative elements that Guarini introduced into the mathematical treatises of the epoch.[35] Dechales limited himself to underlining the order and the arrangement of the propositions with respect to Euclid's text, while Tricomi [1970] based his judgement on modern mathematics. If the *Euclides adauctus* is compared to its *Appendix* and the work which followed, *Modo di misurar le fabriche*, it can be seen how different a treatment there is between an in-depth work of geometry and a manual aimed at instructing a less cultured readership.

As far as originality is concerned, and the consequences of the results obtained by Guarini in the field of mathematics, while it is true that we find only one proof that is somewhat simplified, or alternate ways of proving an assertion, or again generalizations, exemplifications and exercises regarding new solids, but no genuine revolution, we must in any case recognize the exceptional depth and originality in the fact that Guarini was able to insert an extraordinary variety of geometrical shapes in his artistic creations. Mathematics is made visual in the plans of his churches and palaces, in the shape of the stairs, in the interior and exterior decorative forms, in pavements, windows, the shapes of domes, and in hundreds of architectural details, such as curves and polygons, stars, tilings, vaults, friezes and more. Thanks to this artistic wealth and sensitivity, as well as attention to teaching, we are well justified in defining Guarini as the artist-mathematician of the seventeenth century.

Translation from the Italian by Kim Williams

Appendix I: Index of Euclides adauctus et methodicus mathematicaque universalis (Torino 1671)

Regalis Celsitudo
Benevolo Lectori
Imprimatur
Index
Tractatus I Praeliminaris. *De essentia quantitatis continuae :* 1-12.
Tractatus II Praeliminaris. *De essentia quantitatis discretae* : 13-20.
Tractatus III. *De Mathematica eiusque Affectionibus* : 21-32.
Tractatus IV. *In primum Librum Elementorum* : 33-52.
Tractatus V. *In secundum Librum Euclidis de aequipotentia linearum* : 53-62.
Tractatus VI. *In Euclidis Librum tertium de Circulis* : 63-82.
Tractatus VII. *In Librum quartum Elementorum, De inscriptione & circumscriptione figurarum in circulo* : 83-91.
Tractatus VIII. *Arithmetica simplex et generalis integrorum numerorum* : 92-105.
Tractatus IX. *In V Librum Euclidis, Pars prima De Proportionum Notione* : 106-118. *In V Librum Euclidis, Pars secunda De Proportionibus in genere* : 118-131.
Tractatus X. *In VI Librum Euclidis, De proportione quantitatis continuae* : 132-153.

Appendix II: *Index of the* Appendix ad Euclidem adauctum (Torino 1671; 1-24 non-numbered pages)

Appendix II: *Index of the* Modo di misurar le fabriche (Torino 1674)

Notes

1. ...*sed insuper Thaumaturga Mathematicorum miraculorum insigni, vereque Regali architectura coruscat. ... Excipiat itaque R.V.C. pacatu vultu, serenaque clementia illam, quam toties in concipiendis sublimibus idaeis vasto ingenij sui sinu fovit mathesim, et in ea adornanda exanthlatos laboris mei conatus* [Guarini 1671: Regalis Celsitudo, p. 2 (not numbered)].
2. ...*omnes artes vel a Mathematica, vel a Philosophia, vel a Medicina dependent, quae omnes scientiae vel rerum similitudinem, vel proportionem, vel convenientiam considerant.*
    Nam quanto magis artifex abundat in rerum cognitione ad artem suam spectantium, convenientiaeque earum, noverit omnes modos, et multimodum earum applicationem, tanto magis excellens dicitur et perfectius operatur.
    Nam cum artifex vult operari, oportet ut seligat materiam aptam, instrumenta noverit, connexionem rerum ad artificium spectantium et tandem illud artificium unde quaque decreverit. Et quia nec imaginatio, nec intellectus in difficilibus aliquando sufficiunt, hinc est quod modulos parvos artifices conficiant, ad ideam perfectius in ipsa re probandam et perficiendam [Guarini 1665: 213].
3. *Delle operazioni per così dire infinite che i matematici vanno esercitando con evidenti dimostrazioni, ne sceglieremo alcune le più principali, che sono necessarie all'Architettura, senza però arrecare le prove, perchè questo si è proprio uffizio della Matematica, di cui l'Architettura si professa discepola* [Guarini 1736: 18].
4. L. B. Alberti, *De re aedificatoria*, Florentiae 1485; F. di Giorgio Martini, *Trattato di architettura civile e militare*, 1482-92; A. Dürer, *Unterweisung der Messung mit dem Zirckel und Richtscheyt...*, Nurnberg 1525.
5. *Euclides adauctus et methodicus mathematicaque universalis Caroli Emanueli II Sabaudiae duci Pedemontium Principi Regi Cypri etc. dicata, quae ne dum propositionum dependentiam, sed et rerum ordinem observat. Et complectitur eaomnia, quae de quantitate tum discreta, tum continua abstracta speculari queunt. Resectis superfluis demonstrationibus, et requisitis omnibus profuse coadunatis. Singuli quoque Tractatus novis propositionibus*

adaucti sunt, et aliqui etiam exintegro adornati. Omnesque tum figuris, tum verbis clare, dilucideque propositi.

6. *Siquidem ex meo labori didici, cuius pretij, cuius utilitatis id operis emergat,* quod ea omnia quae Mathematicas luces et evidentias in unicum lucis fontem *adeoque solem ne dum tumultuaria collectione aglomeret, sed etiam ordinato agmine disponat in seriesque suas naturali consecutione distinguat praecipue illis qui nullo Mercurio tramitis indice aut duce audent se huic studio consignare et admodum dificilem provinciam in suam sarcinam traducere* [Guarini 1671: *Benevolo Lectori*, p. 1 (not numbered)] (emphasis mine).

7. The first volume of Hérigone's work (1634) contained Euclid's Elements and Data, an appendix on plane geometry, the books of Apollonius on loci and the doctrine on the division of angles; the second (1634), practical arithmetic and algebra; the third (1634), practical geometry, fortifications, mechanics and tables of sines and logarithms; the fourth (1634), geography and navigation; and the fifth (1637), the sciences relative to optics, perspective, Theodosius's spherical trigonometry, the theory of planets, gnomonics and music. As the title notes, the entire encyclopaedia was written in two languages, Latin and French.

8. As the title implies, the corpus of the work, divided into twenty-eight books, is organised in order that even beginners, without an instructor and through their own efforts, can learn all of mathematics starting with the fundamental elements.

9. Reprinted in 1690 with the addition of a new volume that included all of Dechales's handwritten notes made from the publication of the first edition to the time of his death, the work opened with a historic overview regarding progress made in mathematics and presented the most valuable books over the course of thirty-one scientific sections: Euclidean geometry, Theodosius's spherical geometry, the conics, arithmetic, trigonometry, practical geometry, mechanics, statics, geography, magnetism, civil architecture, the art of building in wood, stonecutting, military architecture, hydrostatics, hydrodynamics, hydraulic machines, navigation, optics, perspective, catoptrics, dioptrics, music, pyrotechnics, the astrolabe, the theory and use of sundials, astronomy, theory of planets, meteors, and the calendar.

10. *Oratio habita in funere Reverendi Patris Claudii Francisci Milliet Dechales Societati Iesu, in Collegio Taurinensi eiusdem Societatis die 28 Martij 1678*, Taurini, Typis B. Zapatae, 1678.

11. *Unde in iura Mathematica maxime illi peccant, qui, ut ex ungue discamus Leonem, dum unam propositionem probant, alias quae illius loci non sunt, praxes propositionesque ex aliis non cognitis ostensas, aut tantummodo assertas adferunt et sic mentes discentium tenebris offundunt et in ambages urgent* [Guarini 1671: *De Mathematica instructione*, 25].

12. [Guarini 1671: *De quantitate continua*, 1-12; *De quantitate discreta*, 13-20; *De Mathematica ejusque affectionibus*, 21-32]. See also [Guarini 1665: *De quantitate*, 118-120; *De continui compositione*, 249-266].

13. This discussion also regarded infinitesimal magnitudes for which the theory of proportions did not hold.

14. Cf. [Guarini 1671: Exp. iv, pp. 5-7, *Puncta infinita in quantitate, an admitti debeant* ("if infinite points must be admitted in quantities")].

15. *Bonaventura Cavallerius per indivisibilia libro ad id conscriptum non sine ingenio et subtilitate Mathematicam se promovere profitetur et ex contemplatione punctorum indivisibilium in quantis existentium aequalitates et proportiones Mathematicorum corporum invenire* [Guarini 1671: 11].

16. Mario Bettini (1582-1657), a Jesuit from Bologna, taught mathematical philosophy and moral philosophy at the Gymnaseum in Parma. Here Guarini is referring to vol. 2 of his *Aerarium Philosophiae Mathematicae*, published in 1648, in which he confutes the doctrine of indivisibles in the *Epilogus Planimetricus*, Pars II, § XX-XXII, [Bettini 1647-48: vol. 2, Pars II, 24-37].

17. [Bullialdus 1657 : Prop. XLII, Nota II, 66-67]. Guarini probably consulted the work of Ismaël Bullialdus (1605-1694) during his sojourn in France. In his essay on spirals, Bullialdus praises Cavalieri, although he does mention the criticisms of his contemporaries regarding indivisibles.

18. Cf. [Guarini 1671: 21-32]. G. Vitali's *Lexicon Mathematicum* [1668] is referred to in the definitions of sine (*sinus rectus*), cone, the object of geodesy, and the history of the problem of squaring the circle; cf. [Guarini 1671: 307, 390, 503, 527].

19. *Mathematica in tres partes dividi potest in Mathematicam Universalem, Cosmicam et Microcosmicam. Prima est duplex, nam alia est quae agit de quantitate discreta, alia de continua. Secunda quoque duplex est, alia agit de coelo, altera agit de terra, terrenisque omnibus quae mensuris subsunt. Tertia quoque duplex est. Alia enim pertinet ad naturam hominis, ut visus circa quem versatur Optica. Alia spectat ad artem, ut Mechanica et hujusmodi. Hoc autem libro tradimus Mathematicam Universalem, quae de omni quantitate in communi peragit et omnibus aliis mathematicis partibus aditum aperit* (Mathematics can be divided into three parts: universal, which in its turn is divided into two parts, one dealing with discrete quantity and the other with continuous; cosmic which deals with the world, that is, with the heavens, earth, and all earthly things subject to measure; microcosmic which deals with human nature and its natural activities such as optics or those relative to the applied arts, such as mechanics, etc. In this work however we will deal with universal mathematics that generally regards all quantities and opens the way to all of the other parts of mathematics) [Guarini 1671:23].

20. [Clavius 1574], cited in [Guarini 1671: 3, 25, 46, 67, 78, 91, 131, 145, 176].

21. [Commandino 1572], cited in [Guarini 1671: 54, 133].

22. [Tartaglia 1565], cited in [Guarini 1671: 35].

23. From the book *Tabula tetragonica, seu quadratorum numerorum* by Paduan mathematician G. A. Magini, who taught at Bologna, Guarini took the representation of the plane numbers designed with little stars; cf. [Magini 1592: 1] and [Guarini 1671: 219, 222].

24. [Bettini 1647-48: 183-210] cited in [Guarini 1671: 249].

25. [Clavius 1603] cited in [Guarini 1671: 222, 293].

26. [Benedetti 1563] cited in [Guarini 1671: 85].

27. [Guarini 1671: *In Librum quartum Elementorum De inscriptione et circumscriptione figurarum in circulo*, 83-91]. This is how he underlines the importance of this topic for artists and artisans: "the fourth book [of Euclid's *Elements*] deals with the construction of figures with respect to the circle, however, it is more convenient, with the other polygons, to execute the construction with the inscription or circumscription of the circumference. The use of this book is absolutely required for artisans and artists, both for the solids which have to be inscribed in the sphere or must circumscribe this, as well as for determining the relationship between an outer solid polyhedron and an inner one. We recall that with a procedure of this kind Archimedes found the volume of the sphere. Such constructions are also useful for determining the lines and chords of arches, and also for laying out the drawings of military fortresses" [Guarini 1671: 83].

28. Cf. [Clavius 1612: vol. 1 (*Euclidis Elementa*), Lib. 6, 296-300; vol. 2 (*Geometria practica*), Lib. 7, Appendix, 189-192] and [Guarini 1671, Exp. IV *De linea quadratrice*, pp. 293-296]. In addition to the construction by Clavius, he includes the one given by the Jesuit Vincent Leotaud in his *Cyclomantia seu de multiplici circuli contemplatione libri 3* (Lyon, 1663), which is also cited in *Architettura civile* [Guarini 1968: tratt. I, Cap. X, Oss. VI, 59].

29. [Clavius 1612: vol. 2, Lib. 8, prop. 47, 218-219] and [Guarini 1671: *De circuli segmentis in figuram circularem coaptandis*, 289; Probl. 2, Prop. 7, *Lineam ovalem proprie dictam efformare*, 290]. Here Guarini uses the construction of the mean proportional, already described in treatise XV [Guarini 1671: 248-249], that is, the proportional compass that is already found at the beginning of René Descartes's *Géométrie*. The construction recalls that given by Teofilo Bruno, a mathematician from Verona in [Bruno 1627: 2-4], and [Bruno 1631: 72-73]. For more on this, see [Ulivi 1990].

30. [Clavius 1612, vol. 4 (*Gnomonices*), 28-30, 75-78]; [Guarini 1671: XVIII, Exp. VI *De linea ellipsi*, 297]. Guarini defines the ellipse kinematically: the motion of a point that goes further away from one focus as it goes nearer to the other. He then states that it is also obtained as a section of a cone and presents the classical gardener's construction.

31. [Benedetti 1574: 39]. Cf. also [Benedetti 1585: 348-351] on the instrument he conceived for tracing the curve.

32. *Ambrosium Vincentium virum in Mathematicis admirabilem, in quo quaedam etiam desumemus in sequentibus* (Ambrosio Vincenzo, admirable expert in mathematics, from whose

work we have gathered some of the results which follow) [Guarini 1671: 419]; cf. also [Guarini 1671: 420, 495].

33. *Appendix ad Euclidem adauctum Guarini Guarinii c.r. Theatini. Quoniam multa, quae ad cubationem corporum faciunt, quae a nemine tacta; et animadversa sunt, mihi post impressionem libri occurrerunt, quae ne dum erant utilia, sed pene necessaria, & stereometria practica deficiebant maxime concamerationum quadratarum cuiuscunque generis, volusiis plurimus corporibus, quae cubationi subieci in nostro hoc opere, haec omnino illis subnectere, ut iam nullum sit corpus sub aliqua certa superficie comprehensum quod corporum cuborum mensuris non sit subactum, & mathematica certitudine illius mensura poenitus non innotescant.* In the copy of Guarini's *Euclides adauctus* conserved in the Biblioteca Nazionale Universitaria in Torino, indexed as Cav. 60, this *Appendix* has been bound at the end of the volume.

34. Rhomboid-solids are bodies made of two right cones which have bases that are equal and adjacent. When the rhomboid solid is cut with a plane passing through the axis and perpendicular to the bases of the cones, the result is a plane rhombus. In this treatise reference is sometimes made to a rhomboid solid by simply using the term "rhombus."

35. After having cited the contents of the thirty-five treatises of Guarini's *Euclides adauctus*, Dechales writes: *quamvis in hoc opere multa sint optima, methodus tamen, et ordo non arridet, multa item non satis clare explicat. Unde melius scripsisset si Euclidem in suo ordine reliquisset peculiaribusque tractatibus caeteras materias explicuisset. Hic enim ordo confusionem patit* (although there are many excellent things in this work, however the method and the order are not pleasing, and many things are not clearly explained. It would be better if he had left Euclid in his order and had set forth the remaining material in specific treatises) [Dechales 1674: t. I, 27].

Bibliography

ALSTED, J. 1620. *Tractatus de architectura.* Herborn.

ARNHEIM, R. 1977. *The Dynamics of Architectural Form.* Berkeley: University of California Press.

BENEDETTI, G. 1553. *Resolutio omnium Euclidis problematum una tantummodo circini data apertura.* Venice: Apud B. Caesanum.

———. 1574. *De gnomonum umbrarumque ....* Torino.

BERNARDI, M. 1963. *Tre palazzi a Torino.* Torino: Istituto Bancario S. Paolo.

BETTINI, M. 1647-48. *Aerarium Philosophiae Mathematicae,* 2 vols. Bononiae: J. B. Feroni.

BOULEAU, C. 1963. *The Painter's Secret Geometry: A Study of Composition in Art.* New York: Harcourt & Brace.

BRUNO, T. 1627. *De naturali et vero corpore ovato atque eius sectione et formatione.* Vicenza: F. Grossum.

———. 1631. *Dell'Armonia astronomica et geometrica, parte seconda.* Vicenza: Grossi.

BULLIALDUS, I. 1657. *De lineis spiralibus.* Paris : Apud Sebastianum et Gabrielem Cramoisy.

CARAMUEL, J. 1670. *Mathesis biceps vetus et nova.* Campaniae: Officina Episcopale.

CAVALIERI, B. 1632. Lo specchio ustorio. Bologna: C. Ferroni. Rpt. 2001, E. Giusti, ed.

———. 1645. *Geometria indivisibilibus continuorum nova quadam ratione promota.* Bononiae.

CERRI, M. G. 1985. *Architetture tra storia e progetto. Interventi di recupero in Piemonte 1972-1985.* Torino: Allemandi.

———. 1990. *Palazzo Carignano. Tre secoli di idee, progetti e realizzazioni.* Torino: Allemandi.

CERRI, M. G., TURCHI E., CARENA L. 1997. *Un simbolo del Barocco a Torino. La Cappella della Sindone 1610-1997.* Torino: Unesco.

CLAVIUS, C. 1574. *Euclidis Elementorum libri XV.* Rome: Apud V. Accoltum. (Rpt. Rome 1589, Köln 1591, Rome 1603, Köln 1607, Rome 1629, Frankfurt 1654.)

———. 1606. *Geometria practica.* Rome.

———. 1612. *Opera Mathematica,* 5 vols. Moguntiae: R. Eltz.

COMMANDINO, F. 1572. *Euclidis Elementorum libri XV.* Pesaro.

———. 1575. *Degli Elementi d'Euclide Libri quindici con gli Scholii antichi tradotti prima in lingua latina ... e con commentarii illustrati, et hora d'ordine dell'istesso trasportati nella nostra vulgare et da lui riveduti.* Urbino: D. Frisolino.

─────. 1588. *Pappi Mathematicae Collectiones* .... Pesaro.

COXETER, H. S. M. 1961. *Introduction to geometry*. New York: John Wiley & Sons.

─────. 1963. *Regolar polytopes*. New York: Macmillan.

D'AGUILLON, F. 1613. *Opticorum libri sex Philosophis iuxta ac Mathematicis utiles*. Antwerp: Ex of. Plantiniana.

DECHALES, C. F. M. 1674. *Cursus seu Mundus Mathematicus*. Lyon.

DI MACCO, M., G. ROMANO. 1989. *Diana trionfatrice. L'arte di corte nel Piemonte del Seicento*. Torino: Allemandi.

GUARINI, G. 1665. *Placita Philosophica*. Paris: Dionysium Thierry.

─────. 1671. *Euclides adauctus et methodicus*. Torino: Typis Bartholomaei Zapatae.

─────. 1674. *Modo di misurare le fabriche*. Torino: Per gl'Heredi Gianelli.

─────. 1675. *Compendio della sfera celeste*. Torino: Giorgio Colonna.

─────. 1676. *Trattato di fortificazione*. Torino: Appresso gl'heredi di Carlo Gianelli.

─────. 1678. *Leges temporum et planetarum*. Torino: Ex typ. haeredum Caroli Ianelli.

─────. 1683. *Coelestis mathematicae*. Milan: Ex typ. Ludovici Montiate.

─────. 1686. *I dissegni d'architettura civile et ecclesiastica*. Rpt. 1966, D. De Bernardi Ferrero, ed. Torino: Bottega d'Erasmo.

─────. 1737. *Architettura civile*. Torino: G. Mairesseo. Rpt. 1968, Milan: Il Polifilo.

GULDIN, P. 1642. *De centro gravitatis* ... Viennae: Formis Mattaei Cosmerovij, 4 vols., 1635 - 1641.

HÉRIGONE P. 1634-37. *Cursus Mathematicus, nova, brevi, et clara methodo demonstratus, per Notas reales et universales, citra usum cujuscunque idiomatis, intellectu faciles. Cours mathematique, demonstré d'une nouvelle, briefue et claire methode, par Notes reelles et universelles, qui peuvent estre entendüs facilement sans l'usage d'aucune langue*, 5 vols. Paris: H. le Gras.

KLAIBER, S. E. 1993. Guarino Guarini's Theatine Architecture. Ph.D. thesis, Columbia University.

LEOTAUD, V. 1663. *Cyclomantia seu de multiplici circuli contemplatione libri 3*. Lyon.

MAGINI, G. A. 1592. *Tabula tetragonica, seu quadratorum numerorum*. Venice: Ciotti.

MILLON, H. A. 1999. *I trionfi del Barocco. Architettura in Europa 1600-1750*. Milan: Bompiani.

MYDORGE, C. 1639. *Prodromi catoptricorum et dioptricorum sive conicorum operis ad abdita radii reflexi et refracti*, 2nd ed. Paris: Ex typ. Dedin. (1st ed. 1631, Paris).

PACIOLI, L. 1494 *Summa de Arithmetica, Geometria Proportioni et Proportionalità*. Vinegia:P. de Paganini.

─────. 1509 *Divina proportione*, Venetiis: P. Paganinus.

PASSANTI, M. 1941. *La real cappella della S. Sindone in Torino*. Torino: Accame.

─────. 1963. *Nel mondo magico di Guarino Guarini*, Torino: Toso.

[Proceedings 1970]. *Guarino Guarini e l'internazionalità del Barocco*. Atti del Convegno Internazionale Accademia delle Scienze di Torino, 1968. 2 vols. Torino: Accademia delle Scienze.

ROBISON, E. C. 1985. Guarino Guarini's Church of San Lorenzo in Turin. Ph.D. thesis, Cornell University.

─────. 1991. Optics and Mathematics in the Domed Churches of Guarino Guarini. *Journal of the Society Architectural Historians* 50, 4 (December 1991): 384-401.

ROERO, C. S. 1999. Mean, Proportion and Symmetry in Greek and Renaissance Art. *Symmetry: Culture and Science, The Quarterly of the International Society for the Interdisciplinary Study of Symmetry (ISIS-Symmetry)*, Special Issue*: Chapters from the History of Symmetry edited by György Darvas* 9, 1-2: 17-47.

─────. 2000. Media, proporzione e simmetria nella matematica e nell'arte da Policleto a Dürer. Pp. 40-59 in *Conferenze e seminari 1999-2000*, E. Gallo, L. Giacardi, C. S. Roero, eds. Torino: Associazione Subalpina Mathesis.

─────. 2005. *Les symétries admirables de Guarino Guarini*. Pp. 425-442 in *Symétries, Contribution au séminaire de Han-sur-Lesse, septembre 2002*, P. Radelet de Grave, ed. *Réminiscences* 7. Turnhout: Brepols.

————. 2006. La geometria del compasso fisso nella matematica e nell'arte. Pp. 247-274 in *Matematica Arte e Tecnica nella Storia*, L. Giacardi, C. S. Roero, eds. Torino: Kim Williams Books.

ROMANO, G., ed. 1999. *Torino 1675-1699. Strategie e conflitti del Barocco*. Torino: Cassa di Risparmio di Torino.

SAINT VINCENT, Grégoire. 1647. *Opus geometricum quadraturae circuli et sectionum coni*. 2 vols. Antwerp.

SBACCHI, M. 2001. Euclidism and Theory of Architecture, *Nexus Network Journal* 3, 2: 25-38.

SCHOTT, C. 1661. *Cursus Mathematicus sive Absoluta omnium Mathematicarum Disciplinarum Encyclopaedia, in Libros XXVIII digesta, eoque ordine disposita, ut quivis, vel mediocri praeditus ingenio, totam Mathesin a primis fundamentis propio Marte addiscere possit. Opus desideratum diu, promissum a multis, a non paucis tentatum, a nullo numeris omnibus absolutum*. Herbipolis (Würzburg): Jobus Hertz.

————. 1662a. *Mathesis Caesarea*, Herbipolis (Würzburg): Schonwetter, 1662.

————. 1662b. *Physica curiosa sive mirabilia naturae et artis*. Herbipolis (Würzburg): Jobus Hertz.

TARTAGLIA, N. 1565. *Euclide Megarense philosopho, solo introduttore delle scientie mathematice diligentemente rassettato et alla integrità ridotto*. Venice: Appresso Curtio Troiano.

TORRETTA, G. 1968. *Un'analisi della cappella di S. Lorenzo di Guarino Guarini*. Torino: Edizioni Quaderni di studio.

TRICOMI, F. G. 1970. Guarini matematico. Pp. 551-557 in vol. II of *Guarino Guarini e l'internazionalità del Barocco*. Atti del Convegno Internazionale Accademia delle Scienze di Torino, 1968. Torino: Accademia delle Scienze

ULIVI, E. 1990. Il tracciamento delle curve prima di Descartes. Pp. 517-541 in *Descartes: il metodo e i saggi*, G. Belgioioso, ed. Rome: Istituto della Enciclopedia Italiana.

VALERIO, L. 1661. *De centro gravitatis solidorum libri tres*. Bononiae: Ex. typ. Haeredum de Duccijs.

VITALI, G. 1668. *Lexicon mathematicum, astronomicum, geometricum*. 1st ed. Paris: Ex Officina Ludovic Billaine. 2nd ed. Romae, 1690.

VIVIANI, V. 1659. *De maximis et minimis geometrica diuinatio in quintum Conicorum Apollonii Pergaei adhuc desideratum ... liber primus[-secundus]*. Florentiae: apud Ioseph Cocchini.

WILLIAMS, K., ed. 1996. *Nexus: Architecture and mathematics*. Fucecchio: Edizioni dell'Erba.

About the author

Clara Silvia Roero is full professor of History of Mathematics at the University of Torino. She is a member of the editorial board of several journals, including *Bollettino di Storia delle Scienze Matematiche, Revue d'histoire des mathematiques, Lettera Matematica Pristem*, and *Il Maurolico*. She is on the Scientific Committee for the collected scientific papers of the mathematicians and physicists of the Bernoulli family, for the National Edition of the R. G. Boscovich's works, and for the papers of M. G. Agnesi. She was President of the Italian Society of History of Mathematics (SISM) from 2000 to 2008, and a member of the International Commission of History of Mathematics. She is currently Director of the Torino Research Group on History of Mathematics. She was been visiting professor at the Utrecht University Department of Mathematics (1995) and at the Institut de Physique théorique, Université de Louvain-la-Neuve. She has been an invited speaker in several International and National Conferences, including Nexus 2006 in Genova. She is the author of several articles and books on the history of mathematics. Her research topics include: mathematics and art in Greece and the Renaissance; Egyptian mathematics, Zeno's paradoxes, Islamic algebra, G. Benedetti, the history of probability theory, the history of the Leibnizian infinitesimal calculus, Lagrange, eighteenth- and nineteenth-century mathematics at the University of Torino, and the works of G. Peano and his disciples.

Michele Sbacchi

Università di Palermo
Dipartimento Storia e Progetto
nell'Architettura
Palazzo Larderia
C.so Vittorio Emanuele 188
sbacchi@architettura.unipa.it

Keywords: Guarino Guarini,
Girard Desargues, projective
geometry, projective plane,
Baroque architecture, conic
sections, anamorphism,
perspective theory

Research

Projective Architecture

Abstract. Michele Sbacchi investigates the real influence of the notion of projection on architectural design before and during the age of Guarini. He takes into consideration concepts such as light and shadow, abstract line, plane, section, projective geometry and perspective. To do this he looks at the ideas of Gregorius Saint Vincent, Alberti, Guarini, Desargues and de l'Orme, among others.

Introduction

In his *Problema Austriacum* of 1648, Gregorius Saint Vincent, a rather famous Jesuit and mathematician, gives an extraordinary allegory of projection. The frontispiece of his book displays a projection of sunlight which, although it passes through a square body, projects, in fact, as a circle on the ground (fig. 1). On a ray of light appears the motto: *Mutat quadrata rotundis.*

Then in the preface he further clarifies his mythological view:

> *Nihil in humano stabile, nec raro Dominos mutant orbis. Ut traiectos per quadrum radios in orbem deduces Quadrata rotundis mutat Sol, ita prosper adversis...* [1647: III].

Gregorius's life-long scientific concern was the problem of squaring the circle. His coupling this problem with the fascinating issue of projection provides exceptional evidence of the deeply symbolic meaning that projections had, especially in the seventeenth century. Gregorius explained the "miracle" of *quadratura* as a divine projection; the anamorphists during that very epoch dwelled upon the same theme: clearly both were fascinated by the "transformation" achieved through projection.

The verb "to project" comes from the Latin *proicere,* literally "to throw forth." The word forth (*pro*), suggests that the idea of future, as the temporal realm of these operations, is, in some way, involved.

The architectural relevance is manifest: architectural "projects," as such, and "projections," as understood in the terminology of architectural drawing, are both the domain of architects. Whereas in English the common root of "project" and "projection" is partly lost because the term "project" is often replaced by "design", it survives in most Latin languages. In French, for instance, *projet* and *projection* are the relevant terms. Despite the linguistic differences,[1] the idea of projection clearly lurks behind both these words, but whereas it is immediately conjured up in the use of the word "projection", it is, in fact, forgotten in the common understanding and use of the term "project."

In this article I shall try to investigate what the real influence of the notion of projection is, both in literal and metaphorical terms, on architectural design. If we want to make a pun, my aim is to see how "projective" a project is.[2] I will also briefly consider the importance of the projective plane, as the paper-made realm of architectural manipulations.

Nexus Network Journal 11 (2009) 441–454
DOI 10.1007/s00004-009-00011-y; *published online* 7 November 2009
© 2009 Kim Williams Books, Turin

Fig. 1. Gregorius Saint Vincent, *Problema Austriacum* [1648], frontispiece

Light and shadow

The close relationship between the idea of projection and the idea of drawing is paradigmatically fixed in the legend about the origin of painting reported by Pliny.[3] He tells the story of a Corinthian maiden who, on the occasion of the departure of her lover, wanted something to remember him by. She illuminated his face with a lamp and traced the profile of his shadow on the wall. Drawing, then, was born as the marking edge between shadow and light; moreover, it came out of a projection. Quintilian also refers to the birth of painting as the primordial act of drawing around cast shadows.[4]

Apart from the mythological realm, the light/shadow dichotomy has an important place in art and architectural theories. I might quote Alberti who, almost literally in keeping with Pliny's legend, made *circumscriptio* and *receptio luminis* two of the three basic principles of painting in his *De Pictura* [bk. II, 31]. It is also well known that Daniele Barbaro translated the Vitruvian term *Scenographia* as *Sciographia*, rather than "Perspective," thus making "shadowed drawing" one of the three basic forms of architectural representation.[5] No less interesting is the way by which Henrich Füssli built an evolutionary theory of art entirely grounded on the idea of shadows. For him art evolved from shadow-like images (*sciagrammi*) towards more complex forms to reach maturity with fully colored forms of art. The same concept was adopted by Thomas Kirk and by many others and constituted the basis for the triumph of polychrome architecture [Füssli 1801: 10; Middleton 1985]. At the very root of art, then, we find "projection" as a primary act.

Interest in shadows was a widespread phenomenon during the sixteenth and seventeenth centuries even beyond art theories. Studies on both astronomy and perspective had to refer back to the projection of shadows to establish their internal laws. The writer who placed the greatest emphasis on the importance of the projections of shadows was Biagio Pelacani da Parma who combined shadows (previously reserved to astronomy) with optics (see [Da Costa Kaufmann 1975: 266]). In this regard, the name of Giordano Bruno inevitably comes to mind as the common ontological reference for all these diverse interests. The disciplines involved with problems of graphic representation thus focussed on shadows. This is hardly surprising if we think that shadows and mirrored images are the only two forms of "natural representation":[6] they are respectively the products of the two opposite optical phenomena of total reflection and total absorption of light rays and thus the two ways by which nature can duplicate and represent itself without human intervention.[7] In light of these considerations it is easier to understand that Pliny's legend stigmatizes the edge and the passage from natural representation to man-made representation: from shadow to outline. It is a legend of origin: it portrays the very first step of human representation. From it a fully man-made representation will develop – a representation where "projection" happens on designated surfaces.

Abstract line and plane

If we abandon the archaic realm we have dealt with so far, we note that, in our understanding of drawing, the section of the visual rays – the Albertian *intersecatio* – can hardly be regarded as real or natural; it is, in fact, a highly abstract and artificial operation, where the "projective plane" takes the place of the material surfaces, which constituted the support for primordial drawing. A virtual and abstract mental idea substitutes a material object. Of course such a thing was impossible in the Arcadian gracefulness of Pliny's scene, but this was due to a peculiar reason: in that realm, drawing was identified with a real line. It describes the verisimilar outline of the object. It has, therefore a strict link with reality. Lines, however – and this is the key – can, in varying degrees, also be abstract.

Another legend can help us to grasp the notion of abstract line. It is the legend reported by Diogenes Laertius and Plutarch, and it ties the discovery of abstract lines to the name of Thales and his attempt to measure the Pyramids.[8] As the story goes, Thales ingeniously thought of comparing the height of the Pyramid and the length of the shadow cast with those of a vertical object such as a stick: the shadow of the Pyramid and that of the stick would be proportional. He therefore constructed two virtual triangles out of the vertical axes of the objects, the lengths of their shadows and the line linking the apex of the objects and the extremes of the shadows. Through this simple observation Thales solved a problem that had long challenged the ancients: how to measure an unreachable object. The device that he employed is his famous theorem on the similar triangles that still carries his name. Yet the legend is important for another reason: Thales conceived of lines in abstract terms. His two triangles contain projective, "virtual" lines such as the one linking the apex of the pyramid to the end of its shadow cast on the ground. Thus he inaugurated the geometry of abstract lines.[9]

With the conception of abstract line we can move more easily towards that idea of projective plane that I suggested earlier. Yet another notion is necessary: that of "section". It too is significantly missing in Pliny's tale or in the tradition of *skiagraphia*. We note that, in that natural context, light rays are not sectioned, they are merely interrupted by the presence either of an object or a human body. In fact, the notion of abstract line implies, consequently, the rather more elaborate idea of virtual abstract plane. The virtual plane, in turn, permits us to reconsider the somewhat simple process of shadow casting, in the form in which we have been dealing with so far. For a virtual plane, in contrast to an object, allows a simple but fundamental operation for architectural drawing: the cutting of light rays. The idea of section thus comes to life. "Natural" projection and "artificial" section complete so the apparatus of graphic representation.

The "place" of projection becomes, then, an ideal surface, no longer necessarily a material object. It will be the transparent surface suggested by Alberti: ...*non altrimenti che essa fosse di vetro translucente*...(*De Pictura*, bk. I, ch. 12).

The role of the projective plane within the making of architectural design is interestingly emphasized by Alberti and also deeply linked to his notion of composition. As we know, Alberti does not use the term *compositio* in *De re aedificatoria*, the closest term being the rather different one of *concinnitas*: for him, *concinnitas* intrinsically regulates architecture whereas ornaments belong to a complementary beauty. The significant absence of the term *compositio* together with a statement contained in *De Pictura*, substantiate his position. In *De Pictura* Alberti states that architects borrow ornaments from painters, who were entitled to deal with *compositio*: "the architect took from the painter architraves, capitals, bases, columns and all the other fine features of building."[10] Alberti therefore does not neglect the notion of composition, but he relegates it to the two-dimensional realm of pictorial representation. Furthermore he suggests a borrowing procedure from the two-dimensional plane of painters to the three-dimensional reality of architecture. As Hubert Damish has written: "The kinship between the notions of composition and concinnitas is thus established: both refer to the way in which the parts of the same body, of the same object relate and match, But while a body belongs to a three-dimensional space, the notion of composition is valid only insofar as it is ascribed to the two-dimensional projective plane."[11] *Concinnitas* is for bodies what *compositio* is for their projections. The idea of projection is yet present elsewhere in Alberti's theory, as in his parallel of projected ornaments with alphabetic letters. In this case the procedure is different, since the projection is used as far as the outline is concerned. Rather than establishing a hierarchy of disciplines or a literal procedure of borrowing between them, Alberti was seeking the realm, in a very strict sense, of architectural design. Quite clearly, for him, the projective plane assumes a basic role, for it is the *locus* of architecture as the textual space is the *locus* of writing, "For Alberti no architecture is possible if not born on paper, through the function of projection and transcription that the drawing assumes."[12] Paper, then, nothing but paper, is, ultimately, the way of being and the material status of an architectural project.

Not insignificantly Guarini, the most "projective" of the architects, also emphasized the role of paper as the primary support of architectural design to the point of almost identifying the two things. Several times throughout his treatise he recalls the paper-like essence of architectural drawing:

> Drawing, or idea according to Vitruvius, has three parts, the first of which is called Ichnografia, which is a description and expression *on paper* of what will be occupied by the building, which is drawn in plan; the second is called Orthografia or Elevation, and it is the description and expression *on paper* of the elevation of one of its sides"... .[13]
>
> ...Ichnografia being a description *on paper* of buildings... .[14]

He also points out the somewhat crude fact that, if an actual product should be attributed to architects, it will be the "paper of the project," rather than buildings: "the architect does not build walls, nor roofs, nor machines, nor statues, nor doors, nor locks, nor bricks."[15] He consistently regards the production of drawings on paper as the basic activity of architects. This awareness brought him to devote a considerable part of his treatise to topics such as drawing instruments, the use of paper, the making of ink, etc. He says quite explicitly: "the instruments used by Architecture proper in order to direct the Arts subject to it are few because they are only those which serve to draw and represent its ideas *on paper*."[16] Lomazzo, not concerned with building practise, made this process more magical and "envisaged the form emerging from paper as a ghost materializing."[17] It is probably worth going back to the emphasis Alberti places on the idea of "ornaments composed through projection". Alberti's conception was represented, with an interesting shift, by Claude Bragdon in a booklet, which carries the meaningful title of *Projective Ornament* [Bragdon 1915]. Bragdon's "projective ornaments" are superficial decorative patterns composed in a two-dimensional plane. Acting in a post-projective geometry epoch, Bragdon was concerned with exploiting the possibilities of projections for his decorative purposes. The *projectivity* of Bragdon's ornaments lays with the way in which they are begotten – i.e., by means of a broad use of projective geometry. The rule of *compositio*, rendered even stronger by the possibilities opened up by projective geometry, returns. "Architecture composes through painting" as in Hubert Damish's reading of Alberti (and is reinforced by projective geometry, we could add, thus acknowledging Bragdon).

Projective geometry

Bragdon's use of projective geometry is interesting in its peculiarity, since projective geometry, somewhat strangely, even paradoxically, is taken as a compositional device (in a fine way, I would add). Projective geometry was in fact meant to follow rather different paths, almost avoiding uses such as Bragdon's. It was conceived by Gérard Desargues as a deliberate attempt to create a universal rule – a *maniere universelle* as Desargues's very title indicates – applicable to the different realms in which graphic representation was required. Clear evidence of this intention is, paradoxically, the very difficulty that master masons found in using it, because of its extreme abstruseness.[18] Desargues, and his pupil Abraham Bosse, a painter who devoted himself to defending and promulgating his master's theories, took the discipline as a totally neutral tool which, then, could transcend the specificity of the different disciplines.[19] It is significant that currently projective geometry is often contrasted to metric geometry, highlighting the fact that its peculiarity consists in allowing planar transformations which are valid despite numerical attributes.[20] Quite symptomatic of Desargues's and Bosse's rational attitude is their treatment of projections on irregular surfaces (see [Baltrusaitis 1969: 71] (fig. 2).

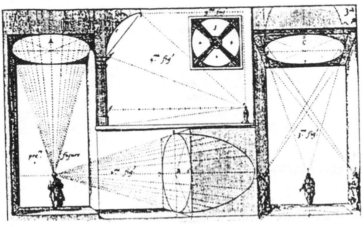

Fig. 2. The rational treatment of perspective from *Moyen universel de pratiquer la perspective sur les tableaux, ou les surfaces irregulieres* [Bosse 1653]

Desargues dedicated an entire treatise to this topic [Bosse 1653] but the variations that the curve or inclined surface of a vault produce on representation are absolutely emptied of their mystic and symbolic power. Whereas Niceron, Scott and many others were fascinated by these irregularities and made them the object of their anamorphic art, Desargues and Bosse did just the opposite, treating distorted representations as ordinary, plain cases. The contrast between the highly symbolic way of dealing of the anamorphists and the cool rational attitude of Desargues and Bosse is even more striking if we consider that they developed their disciplines during the same years, and that both Desargues and Bosse were very close to people like Niceron and the circle of the Minims which was a centre point of anamorphic art.

But it must be borne in mind that the idea of projection went, in that period, through a major change, which, although it originated within the realm of graphic representation would have, as I will try to show, a substantial architectural twist. Up to the Renaissance, perspective was essentially intended to reproduce the natural process of human vision: the use of the same term of *perspectiva* for both optics and perspective is symptomatic of this fact. Similarity, or, better, verisimilitude had firstly to be sought. Consistently, the symmetrical correspondence between the elements of real objects and those of their relevant images was not recognized. Objects, as rendered in perspective, were considered altered (*digradati* was the term frequently used). Attention was indeed paid to what was changed by projection. Even Descartes admitted, with something like regret, that correspondence was not always observed between reality and perspective: "Following the rules of perspective we often better represent circles with ovals and squares with lozenges rather than with other squares ... so that often, to be more perfect as images and to better represent the object, they have not to resemble it at all" [Descartes 1637: 113]. As J. V. Field and Jeremy Gray put it, "Emphasis was then upon what has been changed by the 'projection'" [Field & Gray 1987: 28]. Desargues, differently, introduced a new way of considering the process. His notion of invariance was the root of an understanding in which attention was focussed upon the elements which remained unchanged. This attitude can easily be grasped by looking at one of his few drawings of a human shape contained in Bosse's expanded version of his *Perspective*. But this change cannot be attributed to Desargues's contribution alone. Very probably it was the outcome of developments in several fields and certainly a substantial part was played by stereotomy. Indeed, as we will see later, the theorization of the art of stone-cutting was made on the basis that "projection" implied a symmetrical correspondence, in the full meaning of the word.

Aside from the work already mentioned, Desargues wrote a book on sundials [1640a], a treatise on stereotomy [1636] and a draft for a book on conic sections [1639]. His studies, though original, did not come from an isolated enterprise, but arose from that very prolific scientific community that was the "mathematical" Paris[21] of the seventeenth century. At that time within the Parisian *intellighenzia* there was close contact between scholars of geometry and architects, with a leading role played by Jesuits and Minims, as we have already seen. Desargues's conceptual contribution to projective geometry consists mainly in the notion of "point at infinity" and "line at infinity," to which later Poncelet added the "plane at infinity". These are developments of the idea of vanishing point originally formulated in 1600 by Guido Ubaldo Del Monte as *punctum cuncursus*.[22] The striking contrast between Desargues's mathematization of the notion of "point at infinity" and the still symbolic understanding of his contemporaries can be clearly grasped by looking at Pietro Accolti's *Lo Inganno degli Occhi* in which, as late as in 1625, he is still talking about *Occhio del Sole* (Eye of the Sun) (fig. 3).[23]

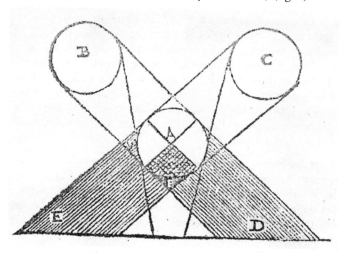

Fig. 3. Projective bands. Pietro Accolti, *Lo Inganno degli Occhi* [1625: 143]

Later, as is well known, the process of rationalization was completed by Monge, who took up these works and systematized them in his descriptive geometry, which, almost unchanged, we still use today. Apart from the merely mathematical notations, what interests me is that projective geometry, in the Desargues-Bosse version, is a discipline both graphically and conceptually centred around the two notions of projection and section, which we have singled out before as foundations of the idea of drawing.[24] But fascination with the possibilities of projections at that time led to a further development of the ancient discipline of conic sections, which is tied to the name of Apollonius. In this regard Desargues is again a key figure: his concern for this subject and his *Brouillon Projet* which, alongside Pascal's *Essay on Conics*, is a seminal contribution in the field, cannot be taken as coincidental. Using again the idea of projection, both Desargues and Pascal at that time visualized the three conics – hyperbola, parabola and ellipse – as projections of the circle. Their studies testify to an uncommon interest paid by French scholars in speculating about the generations of curves, especially when generated by the intersection of different surfaces. Is it likely that architects' experimentations in domes migrated to a broader context? Probably the real common root and stimulus must be sought within the very realm of architecture itself, namely in the vaulting-stereotomy tradition. The making of vaults was, indeed, carried by French architects and master masons to an unbelievably high degree of virtuosity. Vaults and all their manifold derivates such as domes, pendentives, *trompes*, suspended stairs were experimented in the most various shapes, with a spectacular display of unadorned intradoses. This virtuosity gave birth to the more theoretical science of stereotomy – literally "the cutting of solids" – concerned with the possibility of taking control over stone cutting through a two-dimensional representation. The highest possible degree of precision was one of the key requisites. But the real obsession was in the possibility of "making two-dimensional", objects which are three-dimensional, almost making the attempt to unfold them on paper.[25] Stereotomy was a real gymnasium of projections and transformations, in which, as I have already pointed out, the idea of correspondence between drawings and objects was a major concern.

There are, then, enough reasons to think that at that time, especially through Desargues's synthesis, the idea of "projectivity" acted as a common channel linking different disciplines, all more or less on the periphery of architecture. The relationship between Desargues's work on conic sections and his work on perspective has already been demonstrated. Even more striking is the relationship between shadow theory in perspective, conic sections and sun-dialling, this latter being another discipline whose connection with architecture is notoriously sanctioned by Vitruvius's inclusion of it in Book IX of his treatise. The link between these disciplines is so unquestionable that it went, at that time, even to the point of actual confusion: Simon Stevin's book on sun-dialling was translated in Latin as *De Sciagraphia*; John Wells entitled his treatise on sun dialling *Sciagraphia or the Art of Shadows*. Jonas Moore made the connection even more explicit in his "Epistle to the Reader", the introduction to the 1659 English translation of Desargues's work on sun-dials:

> Dyalling I accompt one kind of Perspective, for that glorious Body the Sun, the Eye of the world, traceth out the lines and hour-points by his Diurnal Course, and upon the resubjected Plane by the laws of Picture, Scenographically delineates the Dyal.

He also relates sun-dialling to conic sections:

... this point B [the tip of the gnomon, whose shadow marks the time], you must imagine to be the center of the Earth (for the vast distance of the Sun, maketh the space betwixt the Center and superficies of the Earth to be insensible) and from it at all times of the year the Sun in its course forms two Cones, whose apex is the point B, that next the Sun termed *Conus luminosus* or the light Cone, the other whereof our Author makes use, termed *Conus umbrosus* the dark Cone, now this dark Cone, if by any three points equally distant from the Apex B, the Cone be cut, the section will be a Circle parallell to the Equinoctial: And thereby, as the Author shews many wayes, the position of the Axis or Gnomon may be found out, and the Dyal easily made.

Light, shadow, projection, section: the very ingredients of Pliny's legend are used now in a context in which science and myth are strictly bound together. How these concepts migrated into the very heart of architectural design is clearly shown by the cases of Philibert de l'Orme and Guarino Guarini, not by coincidence both linked to the French milieu. De l'Orme is well known as the first theorist of stereotomy; Guarini's involvement with astronomy, geometry and mathematics is notorious. It is also noteworthy that he taught mathematics in Paris for four years. Yet their concern for disciplines like sun dialling, stereotomy and conic sections went beyond the mere treatment of complementary topics; they were melded into the design method itself. Robin Evans has demonstrated quite clearly how the idea of correspondence and that of projection play a central role in de l'Orme's Chapel of Anet. Guarini is an even more striking example. From his drawings, so full of projections of ceilings or vaults, emerges a deep concern with the architectural "correspondence of elements" between plan and elevation. Yet the idea of projecting one element onto the other is used by him horizontally as well. As Manfredo Tafuri has pointed out, the generation of undulating pilasters in Guarini's church of S. Maria della Divina Provvidenza in Lisbon is the mere consequence of the projection of a twisted column onto the wall: "If a pilaster is nothing but a column projected onto the plane, once chosen the model of a twisted column – present in S. Maria della Divina Provvidenza as a minor order – pilasters will follow their undulated proceeding."[26]

His church of S. Anne La Royale is a further example of "projectivity in plan", and in this case the projection is a central one (fig. 4).

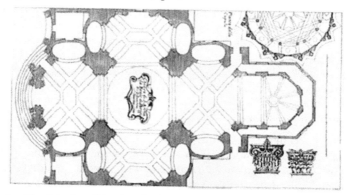

Fig. 4. Guarino Guarini, plan of the church of St Anne La Royale, Paris, *Architettura Civile* [1737]

Guarini will make no mystery of these procedures, For him, quite clearly, a plan is a projection *tout court*: "In projection. therefore Ortografia" [Guarini 1737, bk. IV, I]. And architecture itself, for him, is divided in the somewhat projective categories of *Icnografia*, *Ortografia Elevata* and *Ortografia Gettata*. The definitions of *Ortografia Elevata* and *Ortografia Gettata* are full explications of his design process. On *Ortografia Elevata*:

> The architect has to speculate two kinds of orthography; one which presumes the plane and from it elevates its drawing; a second one which does not presume any drawing in plan, but what is drawn 'above' which has to be later cast onto the plane and see which part is occupied by it; yet orthographies are two, one elevated, one depressed, we will talk about the first in the following treatise...[27]

On *Ortografia Gettata*:

> This orthography is opposed to the previous both by name and by its way of operating; for whereas in the former plane surfaces are elevated by perpendicular lines to give them body and so forming the fabric, this latter on the contrary reduces by perpendicular lines the bodies which are suspended above and reduces them on the plane to unfold their surface...[28]

> ...this is why this Ortografia has been experimented, which indeed puts their surfaces on the plane and shapes them like they are above.[29]

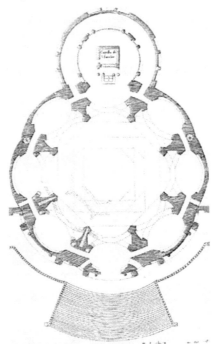

Fig. 5. Guarino Guarini, project for the Sanctuary of Oropa, from *Architettura Civile* [1737]

His dealing with architectural representation is consistent with this idea of "plan surfaces which elevate with perpendicular lines and form the fabric." He has a predilection for drawing in section and especially for constraining in the plan all the elements which will later "explode" in elevation. His plans[30] for the church of S. Lorenzo and for the Oropa sanctuary (fig. 5), in which all the elements of the dome and even its section are projected in one single drawing, are remarkable outcomes of his theory.

Guarini's case is indicative of the notion that the projective dimension of architectural design is not a mere representational constraint. The continuous shifting from the three-dimensional being of architecture and the two-dimensional being of architectural projects – performed constantly during the design process – can strongly orient design choices. The technique of projection, which allows this shifting, makes the 'conceiving of architecture' a highly peculiar process – even from this specific point of view. Therefore, the 'construction of the project' is a process much different from the actual 'making of architecture', this latter taking place in a realm in which the three dimensionality is unquestioned.

Notes

1. I will not deal here with this interesting linguistic shift nor with the rise of the notion of design because this would lead us out of my topic. See [Frampton 1986].
2. I owe the term "projectivity", with this non-mathematical meaning, to Dalibor Vesely.
3. Pliny the Elder (Gaius Plinius Secundus), *Naturalis Historia*, book XXXV; see [Rosenblum 1957].
4. : ... *non esset pictura nisi quae lineas modo extremas umbrae quam corpora in sole fecissent, circumscriberet* [Quintilian, *De Institutione Oratoriae*, bk. X, chap. ii, 7].
5. [Barbaro 1556: 30]; see also Perrault's notorious remark in [Perrault 1979: 10].
6. The development and the conceptual background of outline representation has been masterfully investigated in [Rosenblum 1956]. See also [Ottavi Canina 1982] and [Rykwert 1980: 366ff].
7. It is quite puzzling in this respect that Lomazzo, acknowledging these two extreme positions, divided perspective in *Ottica, Sciographia* and *Specularia*. It is equally puzzling that by the time he wrote the *Trattato della Pittura* Lomazzo was blind. Also interesting is the definition of mirror as "Figuratam, per Esemplare" in the *Vocabolario degli Accademici della Crusca* (Naples, 1747, vol. IV, p. 372). See also [Oechslin 1983].
8. See [Serres 1982 : 85], [Evans 1986] and [Meserve 1983: 222-223]. Less legendary discoveries, made through the same principle, are those of Aristarchus, who made a similar attempt to measure the distances from the earth of the sun and the moon, and of Eratosthenes, who did the same with the circumference of the earth.
9. A distinction has therefore to be made between two types of lines in architectural drawing: those representing actual bodies and those having no corresponding elements in reality.
10. Alberti, *De Pictura*, II, 26 [1972: 60]: *Nam architectus quidem epistilia, capitula, bases, columnas fastigiaque et huiusmodi ceteras omnes aedificiorum laudes, ni fallor, ab ipso tantum pictore sumpsit.*
11. [Damish 1986] ; Savignat draws the same conclusions about Alberti. : "...*la composition de la forme architecturale n'est alors qu'un assemblage de lignes, de figures sur la surface de la feuille de dessin*" [Savignat 1983 : 63].
12. [Damish 1986]. Quite interesting in this regard are Louis Marin's speculations in "Les voies de la carte" [Marin 1982] and in the chapter "Utopiques de la carte" [Marin 1973].
13. *Il Disegno, o Idea secondo Vitruvio, ha tre parti, delle quali la prima dicesi Ichnografia, che è la descrizione, ed espressione* in carta *di quello, che deve occupare la fabbrica, che si disegna nel piano; l'Ortografia, o Alzato chiamasi la seconda, che è la descrizione ed espressione* in carta *della elevazione di una sua Faccia: ...* [Guarini 1737: bk. II, intr.] (emphasis mine).
14. *Essendo la Ichnografia ... una descrizione degli edifici* sulla carta [Guarini 1737: bk. II, intr] (emphasis mine).

15. *l'architett non fabbrica muri, non tetti, non macchine, né statue, né porte, né serrature, né mattoni* [Guarini 1737: bk. I, chap. I, 8].

16. *...gli instrumenti di cui si serve l'Architettura per sé unicamente, in quanta dirige le Arti a sé soggette, son pochi, perché non sono se non quelli i quali servono per disegnare e rappresentare le sue idee* sulla carta [Guarini 1737: bk. I, chap. IV, 21] (emphasis mine).

17. George Hersey [1976: 85] referring to [Lomazzo 1590].

18. Yet it is fair to add that part of the incomprehensibility of Desargues's work is due to the odd botanical language that he adopted. For this, see the letter that Descartes wrote to Desargues to exhort him to use a more accessible language, in [Descartes 1936-1963: vol. III (1940), 228-229]. This fact has been recently considered in [Field & Gray 1987: 60-68].

19. See [Pérez Gómez 1985: 93ff.] and [Scolari 1984: 46].

20. "Projective Geometry: a branch of geometry that deals with the properties of geometric configurations that are unaltered by projective transformation and in which the notion of length does not appear." S.v. "Projective Geometry", *Webster's Third International Dictionary of the English Language* (Springfield, MA: Merriam Webster Inc. Publisher, 1986), p. 1814.

21. I am using this expression after the title of David Smith's booklet, *Historical-Mathematical Paris* (Paris: Les Presses Universitaires de France, 1925).

22. See [Cassina 1961: 306ff]. Meserve instead attributes to Kepler the notion of "point at infinity" [Meserve 1983: 45-47]. Quite correctly Field and Gray, although acknowledging Kepler's intuition, have recognized the conceptual gap between Kepler's and Desargues's concept; see [Field & Gray 1987: 87-89].

23. *... insegnandoci il testimonio del senso visivo manda l'ombre sue parallele al piano ... con la infinita distanza del luminoso degli opachi ... così restiamo capaci potersi all'occhio nostro, in disegnar far rappresentazione di quella precisa veduta di qualsivoglia dato corvo, esposto all'occhio (per così dire) del Sole quale ad esso Sole gli si rappresenta in veduta: onde siccome specolando intendiamo il Sole non vedere giammai alcuna ombra degli opachi, e superficie ch'egli rimiri e illustri ...* [Accolti 1625: 143].

24. See [Young 1930], especially "Introduction, Paragraph 2: Projection and Section, Correspondences."

25. For me it is not, therefore, coincidental that the idea of "beauty produced by precision of execution" has been extolled by a French scientist-architect such as Perrault. It must also be added that the ideology of precision in stereotomy was also brought about by the economic reason of using the material with the least possible amount of waste. For this see [Potié 1984].

26. *Se la lesena, infatti, non è altro che la colonna proiettatta sui piano, una volta scelto il modello della colonna tortile – presente nella Divina Provvidenza come ordine minore – le lesene dovranno seguirne l'andamento ondulato* [Tafuri 1970: 672, note 1]. Tafuri has also pointed out that projections operate in two other works by Guarini: the curved facade of the Annunziata church and the Tabernacle in Verona: *La nuova legge guida è proprio quella proiezione delle gerarchie dello spazio interno sui piano ... La meccanica combinatoria dei corpi geometrici si proieta, qui, sulla struttura discreta della parete inflessa e articolata* [Tafuri 1970: 669].

27. *DELLA ORTOGRAFIA ELEVATA. Due sorte di ortografia deve speculare l'architetto; l'una che presuppone il piano, e da esso solleva il suo disegno; l'altra che non presuppone alcun disegno sui piano, ma quello che si disegna in alto, che poi si deve gettare in piano, e vedere qual parte vien occupata da esso: però due sono le ortografie, una si dirà elevata, l'altra si chiamerà depressa; di questa ne scriveremo nel trattato seguente...* [Guarini 1737, bk. III, intr. (1968: 113)].

28. *DELLA ORTOGRAFIA GETTATA. Questa ortografia siccome è opposta nel suo titolo all'antecedente, così anche nel suo modo di operare; perché là dove in quella le superfizie piane si innalzano con linee perpendicolari, per dare a loco corpo, e formare la fabbrica, questa per lo contrario i corpi in alto sospesi con linee perpendicolari riduce in piano per istendere la loro superficie...* [Guarini 1737, bk. IV, intr. (1968: 288)]. The peculiar expression, *per dare a loro corpo*, to give them body, is noteworthy.

29. ...*perciò è stata ritrovata questa Ortografia, che appunto mette le loro superfizie in piano, e le forma, come sono in alto, e sarebbero nel proprio loro luogo, di questa abbiamo a ragionare.* [Guarini 1737, bk. IV, intr. (1968: 288)].

30. The plan of S. Lorenzo raises the question about Guarini's use of the term *vestigium* to designate a plan. The same term occurs in another plan. Although the attribution of these captions to Guarini himself is quite doubtful, the term remarkably occurs in his treatise: *La Ortographia non è altro, secondo che provo nel nostro Euclide al tratt. 26 che una impressione, terminazione o vestigio notato nel piano di una superficie ad esso normale* [Guarini 1737: bk IV, I]. Aside from the Vitruvian connections, Guarini's *vestigium* is particularly puzzling if confronted with Desargues's argument about the substitution, in architecture, of the term "plan" with that of *assiette*.

References

ACCOLTI, Pietro. *Lo inganno degli occhi*. Firenze, 1625.

ALBERTI, Leon Battista. 1966. *De re aedificatoria* (1486). Latin text and Italian trans. by G. Orlandi. Milan: publisher.

———. 1972. *On Painting and Sculpture. The Latin Texts of De Pictura and De Statua*. Cecil Grayson, ed. & trans. London: Phaidon.

———. 1988. *The Ten Books on Architecture*. J. Rykwert, N. Leach, R. Tavernor, trans. Cambridge, MA: MIT Press.

BALTRUSAITIS, J. 1969. *Les perspectives fauseé. Anamorphoses ou magie artificielle des effects merveilleux*, Paris: Hatchette. (Engl. ed. *Anamorphic Art*, New York, 1977).

BARBARO, Daniele. 1556. *I dieci libri dell'architettura di M. Vitruvio*. Venice.

BERTELLI, Carlo. 1986. La composizione in Leon Battista Alberti: tra pittura e architettura. *Casabella* **520-21** (Jan-Feb. 1986): 52-60.

BOSSE, Abraham. 1643. *La Pratique du trait à preuves de Mr Desargues, Lyonnois, pour fa coupe des pierres en l'Architecture*. Paris.

———. 1648. *Manière universelle de Mr Desargues pour pratiquer la perspective par petit-pied,comme le Geometral*, Paris.

———. 1653. *Moyen universel de pratiquer la perspective sur les tableaux, ou les surfaces irregulieres*, Paris.

BRAGDON, Claude Fayette. 1915. *Projective Ornament*. New York: The Manas Press.

CASSINA, Ugo. 1961. *Dalla geometria egiziana alla matematica moderna*. Rome: Cremonese.

DA COSTA KAUFFMANN, T. 1975. The Perspective of Shadows, the History and Theory of Shadow Projection. *Journal of Warburg and Courtauld Institutes* **XXXVIII**.: 258-287.

DAMISH, Hubert. 1986. *Comporre con la pittura*. *Casabella* **520-21** (Jan-Feb. 1986): 61-63.

DESCARTES, René. 1637. *Dioptrique*. Leyden.

———. 1936-63. *Correspondence*, 8 vols. C. Adam and G. Milhaud, eds. Paris: Presses Universitaires de France.

DESARGUES, Girard. 1636. *Exemple de l'une des manières universelles du S.G.D.L. touchant la pratique de la perspective sans emploier aucun tiers point, de distance ny d'autre nature.*

———. 1639. *Brouillon proiet d'une atteinte aux evenemens des rencontres du Cone avec un Plan*, Paris.

———. 1640(?)a. *Leçons se Tenebres*, Paris. (lost)

———. 1640b. *Brouillon Proiet du S. G. D. L. touchant line manière universelle de poser le style & tracer les lignes d'un Quadran aux rayons du Soleil, en quelqu'oncque endret possible, avec la Reigle, le Compas, l'equiere & le plomb*. Paris.

———. 1659. *Mr De Sargues' Universal Way of Dyalling*. Daniel King, trans. London: Thomas Leach.

EVANS, Robin. 1986. Translation from Drawing to Building. *AA Files* **12**: 3-18.

FIELD, J.V. and J. GRAY. 1987. *The Geometrical Work of G. Desargues*. New York: Springer-Verlag.

FRAMPTON, Kenneth. 1986. Anthropology of Construction. *Casabella* **520-21** (1986): 26-30.

FÜSSLI, Henrich. 1801. *Lectures on Painting*. London.

GREGORIUS SAINT VINCENT. 1647. *Problema austriacum plus ultra quadratura circuli*. Apvd Ioannem et Iacobvm Mevrsios (Antwerp).

GUARINI, Guarino. 1686. *Dissegni di Architettura Civile ed Ecclesiastica inventati e Delineati da G. Guarini*. Torino.

———. 1671. *Euclides adauctus et methodicus matematicaque universalis*. Torino.

———. 1737. *Architettura Civile*. Torino. (rpt. Milan: Il Polifilo, 1968).

HERSEY, George. 1976. *Pythagorean Palaces*. Ithaca, NY: Cornell University Press.

LOMAZZO, Giovanni Paolo. 1584. *Trattato della Pittura*. Milan.

———. 1590. *Idea del tempio della Pittura*. Milan.

MARIN, Louis. 1973. *Utopiques, jeux d'espace*. Paris: Minuit. (Eng. trans. *Utopics. Spatial Play* (London: MacMillan, 1984).

———. 1982. *Cartes et Figures de la Terre*. Exhibition Catalogue. Paris: Centre Georges Pompidou.

MESERVE, Bruce. 1983. *Fundamental Concepts of Geometry*. New York: Dover.

MIDDLETON, Robin. 1985. Perfezione e colore: la policromia nell'architettura francese del XVIII e XIX secolo. *Rassegna* **23**: 55-67.

OECHSLIN, Werner. 1983. La metafora dello specchio. *Rassegna* **13**: 21-27.

OTTAVI CANINA, Anna. 1982. Verso l'astrazione lineare. In *Il settecento e l'antico*, vol. II of *Storia dell'Arte Italiana. Dal Cinquecento all'Ottocento*. Torino: Einaudi.

PÉREZ GÓMEZ, Alberto. 1985. *Architecture and the Crisis of Modern Science*, Cambridge, MA: MIT Press.

PEROUSE DE MONTCLOS, Jean Marie. 1982. *L'Architecture a la française*. Paris.

PERRAULT, Claude. 1979. *Les dix livres d'architecture de Vitruve* (Paris, 1684). Bruxelles: Pierre Mardaga.

POTIE, Philippe. 1984. *Philibert de l'Orme. La Theorie du Projet Architectural a la Renaissance*. Paris : EPHESS.

ROSENBLUM, Robert. 1957. The Origin of Painting: A Problem in the Iconography of Romantic Classicism, *Art Bulletin* **39**: 279-90.

———. 1956. The International Style of 1800: A Study in Linear Abstraction. Ph.D. thesis, New York University.

RYKWERT Joseph. 1980. *The First Moderns*. Cambridge, MA: MIT Press.

SAVIGNAT, J. M. 1983. *Dessin et Architecture*. Paris: Ecole Nationale Superieure des Beaux Arts.

SCOLARI, Massimo. 1984. Elementi per una storia della Axonometria. *Casabella* **500**: 42-49.

SERRES, Michel. Thales au pied des Pyramides. In *Hermes*, vol. II. Paris: Minuit.

———. 1982. Mathematics and Philosophy: What Thales Saw.... In *Hermes: Literature, Science, Philosophy*. Baltimore: Johns Hopkins University Press.

TAFURI, Manfredo. 1970. Retorica e Sperimentalismo: Guarini e la tradizione manierista. Vol. I, pp. 667-704 in *Guarino Guarini e l'internazionalità del Barocco*. Torino: Accademia delle Scienze.

TATON, René. 1951. *L'oeuvre mathematique de G. Desargues*. Paris.

YOUNG, John Wesley. 1930. *Projective Geometry*. Chicago: Open Court Pub.

About the author

Michele Sbacchi is Associate Professor in architectonic and urban composition at the Faculty of Architecture of the University of Palermo. He is also a practicing architect.

Pietro Totaro

Università di Roma "Tor Vergata"
Facoltà di Ingegneria
Dipartimento di Ingegneria Civile
Via Politecnico 1 - 00133 Rome
ITALY
pietro.totaro@fastwebnet.it

Keywords : Guarino Guarini,
Giacomo Del Duca, Baroque
architecture, façade design,
proportional ratios

Research

A Neglected Harbinger of the Triple-Storey Façade of Guarini's Santissima Annunziata in Messina

Abstract. The triple-storey façade, one of the most original inventions of the Baroque period in terms of form and proportion, arose in Sicily and quickly spread throughout Italy, to Europe and beyond. Guarini's design for the façade of Santissima Annunziata in Messina paved the way for its general acceptance. The roots for the concept, however, may be found in the work of Giacomo Del Duca.

Fig. 1. The parish church in Randazzo, Sicily, showing the bell tower

A new triple-storey schema for the church façade, defined both in its proportions and in its intrinsic structure, appeared in the late Sicilian Baroque. This triple-storey façade – with is top storey often used as belfry – is considered one of the most amazing inventions of that period. Extraordinarily fitted to its functions, both practical and symbolic, this architectonic form would soon find widespread acceptance in Central Europe as well as Central and South America. For this reason – although without good grounds – it is sometimes claimed to have arisen from Iberian and/or Austrian-Hungarian origins. However, the exact history leading up to the definition of such a façade, a very common model for the late Baroque churches in the South-East Sicily, is difficult to reconstruct because documents tracing intervening influences have been lost, both because of negligence and because of the recurring seismic catastrophes [Tobriner 1996].

It is a fact that there are bell towers (fig. 1), built in correspondence to the main entrance of churches during the twelfth and thirteenth century, related to a French-Norman tradition (see the "Clochè a mur" structures common in the South of France). But it is the only extant work of Guarino Guarini in Sicily that laid the foundations for all the future development of the triple-storey façade of the late Sicilian Baroque. In fact, with his strong reputation and his writings, Guarini guaranteed the cultural milieu for the architectural experiments of others, above all, those of Rosario Gagliardi.[1] However, Guarini's invention for the façade of the Theatine church of Santissima Annunziata in Messina is in fact unique in his production. To be precise, the design for the earlier (?) façade of S. Maria in Lisbon, a scheme that clearly shows the influence of Borromini, bears some resemblances to the design of the façade of Santissima Annunziata, although such analogies tend to disappear when one attempts to guess its elevation. A certain formal affinity can also be recognized in the later design of the tabernacle in "San

Niccolò" in Verona. Following Harold A. Meek [1988], one can also notice the sign of the telescopic system which is also present in the other, not executed and likely later, Messinian work: the Church of the Padri Somaschi. Notwithstanding these analogies, the Santissima Annunziata façade remains an exception in both the corpus of Guarini's own architecture and in the panorama of his times.[2] Without a doubt, Guarini "invented" the façade to solve problems generated by the context. First, the body of the church is a pre-existing work. Second, an intervening alteration in the direction of the neighbouring street had resulted in axis of the church no longer being orthogonal to the street.[3] Last but not least, both the Theatines and Guarini nursed ambitions of creating a strong competitor to the Jesuits' Church and College (*Primum ac prototypum collegium*) built to a design of the Messinian Natale Masuccio. The Church of the Jesuits was a double dado surmounted by a dome: it stood out because of its mass (see fig. 2). In contrast, in Santissima Annunziata the dome was far from the street and the axis of the church itself was skewed, as mentioned before.

Fig. 2. S. Giovanni Battista and Jesuit College by Natale Masuccio. Engraving by F. Sicuro, 1768

The architectural design of the churches and their placement were obviously instruments of propaganda. With his invention, Guarini solved the problem: in some sense, the third level of the façade was the projection of the dome over a parallel to the street plane.[4] In fact, Guarini rotated the façade of the church, thus concealing the rotation of the major axis and obtaining, at the same time, the space required to insert the octagonal bell tower which is partly hidden behind the façade (see fig. 3). The façade of the nearby college of the Theatines, with its original arrangement of windows, completed the setting: the affinity with the earlier work of Massuccio for the Jesuits cannot be ignored.

Fig. 3. Santissima Annunziata and Theatine College by Guarino Guarini.
Engraving by F. Sicuro, 1768

In my opinion, all these aspects are the main sources of this work of Guarini's. However, I believe that it is possible to underline – unfortunately only by means of inductive analogies – a path linking the façade of Santissima Annunziata with a series of previous attempts dating back to Michelangelo and the Mannerists, a series that starts from Vitruvius as interpreted by Cesariano. If confirmed, this conjecture would underline Guarini's sensitivity to the suggestions coming from the places he visited. Like any other great architect, Guarini is inventive not only because of what he imagined but also because of the hidden paths he unveiled.

During Guarini's stay in Messina, the city was full of cultural and architectural activity. As underlined by Meek, Messina was at the peak of a process which by that time dated back some centuries, a process cut short by the Spanish repression in 1676. The city's leading citizens, for the most part rich entrepreneurs, had pursued a greater and greater independence. Architecture was a medium used by the city Senate to represent civic pride. To this aim, the best artists of the day were summoned, although not all of them answered the call, put off by Messina's reputation as a provincial city in spite of the wide activity of its port.[5] The architectonic activity was further increased by the undertakings of the new religious orders, above all the Jesuits, who opened the first Gymnasium in 1548, which straight afterwards became the University; in 1604 it was located in the building designed by Natale Massuccio.

Notwithstanding some difficulties, especially the provinciality mentioned above, in the second part of the sixteenth century, Messina's Senate was able to commission famous architects to supervise construction of public buildings. All the architects appointed to oversee building sites in the second half of the century were directly influenced by Mannerism and especially by Michelangelo. More precisely, Andrea Calamech was a pupil of Ammanati, while Giovannangelo Montorsoli and Giacomo Del Duca were pupils of Michelangelo himself.

The first urban planning in Sicily since Federico II can be attributed to Montorsoli: it is the planning of the Piazza del Duomo in Messina (1550). Andrea Calamech executed a wide range of sculptures together with important examples of civil architecture, which were lost during the 1908 earthquake. These included the Palazzo Grano, the ground floor windows of which were adorned with a motive that recall the pediment on the third storey of the façade of Santissima Annunziata (figs. 4 and 5). However, Del Duca is the architect most interesting for the object of this paper: Baglione defined him as having a *spirito gagliardo*, a vigorous spirit [Baglione 1649: 54]. He is characterised as popularist, emotional, heterodox, and a follower of Michelangelo. When the classicist tradition came to be criticised, his work sank into oblivion. Yet it is precisely with Del Duca that Mannerism reaches a transition point towards Baroque. Even Borromini was influenced by Del Duca's work.

Fig. 4 (left). Window from Palazzo Grano, by Andrea Calamech, after [Basile 1960]

Fig. 5 (above). Schematic drawing of the third storey of Santissima Annunciata (drawing by the author)

Del Duca collaborated with Michelangelo above all as a sculptor for the tomb of Pope Julius II, but his last collaboration for Porta Pia was the one that most influenced his future architectural activity. Actually, the extent of Del Duca's contribution to the design of Porta Pia is still unclear, although it may be more prominent than usually assumed [Ackerman 1961: 125-126]. However, in the general plan of Porta Pia the influence of the models appearing in Cesariano's edition of Vitruvius's *De Architectura* can be recognized. Painter and architect — we do not know much about his work — Cesariano was the first to publish an integral version of Vitruvius's work and his edition was the most widespread in his times. Cesariano's knowledge of classical architecture was rather approximate, so his reconstruction of Vitruvius's drawings owes more to his imagination and the culture of northern Italy, which was still more attuned to canons of the Middle Ages than to the analysis of Roman architecture. Still, the book's success bestowed the status of canons on Cesariano's hypotheses. Proof of this is the diffusion of

Cesariano's interpretative schemes even in much later times.[6] Among Cesariano's hypotheses there is one for the temple *in antis* for which two alternatives are proposed: the first one is based on the classic tympanum, the second one calls for a minor order superposed on the first order, joined by means of two semi-tympanums – which can be straight or volute shaped – and culminating with a small tympanum (fig. 6).

Fig. 6. Two hypotheses for a temple *in antis* by Cesariano [1521: bk. III, ch. i, 52r]

Such a structure seems reminiscent of late Romanesque forms and is clearly a licence taken on the part of Cesariano. However, this model can be recognized in Porta Pia. I think that it had a role in the subsequent definition of the system of a façade based on three orders, a definition that Guarini conceived a century later, most likely via Giacomo Del Duca. Not a few elements are shared between Cesariano's proposal and the architectonical solution by Michelangelo for the façade of Porta Pia that faces the city. Beside the overall scheme, one may compare the coat of arms in Porta Pia and its decorative apparatus with the sarcophagus depicted by Cesariano, as well as the great *oeil-de-boeuf* in the second order of the temple *in antis* with the disc in the second order of Porta Pia (fig. 7). Later, in his church of Santa Maria in Trivio in Rome, close to the Trevi fountain, Del Duca explicitly declared his debt: one can clearly discern an interpretation of the model by Cesariano (see fig. 8).

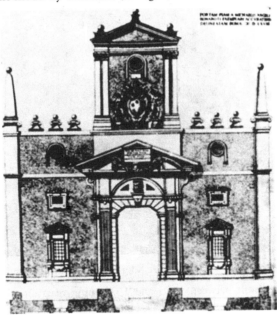

Fig. 7. Porta Pia according to Faleti [Benedetti 1973]

Fig. 8. S. Maria in Trivio, Rome, by Jacopo Del Duca

Fig. 9. The church of the Priory of the Knights of Malta by Giacomo Del Duca

The church was built between 1573 and 1575, nine years after the completion of Porta Pia. Del Duca came back to Sicily for the first time in 1575, although he continued to work in Rome up to 1592, the year in which he was elected to the chair of First Architect of the Messina Senate, at the age of 77. Although he had somehow fallen out of favour in Rome, in Messina the new charge brought him to the apex of his fame. At that time he was given the commission for the modification of the Senate palace and the construction of the headquarters and the church of the Knights of Malta, the latter dedicated to San Giovanni Battista. The church was a demanding project, both because of its renowned commissioners and because its apse was to have housed the chapel in honour of the martyr San Placido and his companions; San Placido's relics had just been found (1588) and the Pope immediately designated him Messina's patron saint. The façade of the *tribuna* clearly displays the results of Del Duca's experience in the work for Porta Pia. He pushed the syntactic decomposition already present in Porta Pia to its extreme (see fig. 9). This only existing urban evidence of Del Duca's work appears as a true façade dramatically marked by strong chiaroscuro effects. It also contains a wealth of stylistic solutions. This led Anthony Blunt [1968] to consider Del Duca's contribution as decisive for the formation of Sicilian Baroque. Remarkable is the use of the giant Doric pilaster, which is hollowed out to "put in evidence" the brick structure. This motif will be used frequently in the Sicilian Baroque from Massuccio's work on. Guarini himself uses it in the façade of the Collegio dei Teatini in Messina. In subsequent years, Guarini used it for the pilasters of the second order in Palazzo Carignano in Torino. This motif is also present in Castellamonte's Piazza San Carlo in Torino (1638), although without the prominence that it enjoys in Del Duca's work. However, it is not likely that Guarini was aware of such works at the time of he was working in Messina. He might, however, have seen the motif in Rome in the works of Del Duca himself, such as for example Villa Rivaldi, the Mattei Chapel at the Aracoeli or in the hint of a pilaster in the rear façade of the Porta San Giovanni. In works of other architects – Alberti's Sant'Andrea in Mantova,

Bramante's San Satiro in Milan, Amadeo's Colleoni Chapel in Bergamo – the pilaster is hollowed out but the recess is essentially aimed to create space for the insertion of a decoration in relief, rarely for the bare treatment we see in Del Duca's style. Exceptions are evident in the frame of some windows, as the those in Palazzo Vidoni-Caffarelli in Rome.[7] The *tribuna* of the Maltese Priory by Del Duca is the only one still standing and, in my opinion, at the time of Guarini would have been sufficient to provoke the interest of the generic observer too. Moreover, it shouldn't be forgotten that Guarini spent the years of his novitiate and architectural education in Rome, at a time when Borromini was active there. The presence of Del Duca was also still discernible in other works of his that could be seen then: Villa Mattei at the Celio (no longer existing), the Orti Farnesiani (today partially destroyed), Porta San Giovanni, the drum, the dome and the bell tower of S. Maria di Loreto, the garden of Villa Bomarzo and that of Villa Farnese in Caprarola. For this reason I think it is likely that Guarini was not unaware of Del Duca's works, and that perhaps he even studied them. Some indications to this effect are present in his work, such as the heads of the cherubs decorating the consoles of the tambour in the church of San Lorenzo in Torino (fig. 10), which are similar to the ones of the ribs-consoles in the final steeple of the lantern in Santa Maria di Loreto (see fig. 11) and those of the Tabernacolo Farnese.

Fig. 10 (left). The cherubs' heads from the drum in the church of San Lorenzo in Torino by Guarini

Fig. 11 (right). The cherubs' heads in the lantern of S. Maria di Loreto by Del Duca

Actually, in the Tabernacolo Farnese the heads decorate the console and do not hold it.[8] Again in Torino, the portal of the stairs that lead to the Chapel of the Holy Shroud form an evident analogy with the window system in Messina's Palazzo Senatorio. Further, in the portal to the Chapel of the Holy Shroud, the two abstract herms on either side of the upper window are reminiscent of Del Duca's peculiar cubic and metamorphic motives.[9] Another of Del Duca's signature stylistic motifs is the keystone that continues in the frame. Such a solution appears in the windows of Santa Maria di Loreto (see figs. 12, 13, 14).

Fig. 12 (left). Guarini's entry portal to the Chapel of the Holy Shroud from the Cathedral, Torino [Meek 1988]

Fig. 13 (right). Engraving of Messina's Palazzo Senatorio by Del Duca

Fig. 14. Window of S. Maria di Loreto by Del Duca [Benedetti 1973]

It is said that in Messina Guarini was in contact with Jesuits, having found in some of them a greater cultural agreement: a possible consequence of this might have been the opportunity for further close examination of Del Duca's architecture. Not only were the

Jesuits Del Duca's mentors, commissioning a number of buildings from him,[10] but they also insisted that he should write a treatise about architecture, although his death prevented its publication [Del Duca 2004].

Nothing is known about the places that Guarini visited during his stay in Sicily. On the basis of the previous observations he might well have visited the places where Del Duca built his works. It is possible for instance that he saw the façade of the Parish Church of S. Agata in Alì, a small town atop the hills near Messina where important families of the Messinian aristocracy spent their summers (fig.15).

Fig. 15. Façade of the Church of S. Agata in Alì

This façade has been repeatedly attributed to Del Duca, although there aren't documents testifying to this.[11] It is known that that construction work had begun on it in 1582 and that the church was almost finished in 1641. In my opinion, one detail not taken into consideration – at least to the best of my knowledge – up until today can be useful to support the attribution of the church to Del Duca. As mentioned earlier, in 1588 the relics of San Placido and his companions were recovered, and Del Duca was commission to build the *tribuna* of the church of the Maltese Priory to house their relics. During the construction, the relics were temporarily housed in the Mother Church of Alì, 25 kilometres away from Messina. It is highly unlikely that at that time this particular a solution would be accepted over other, simpler solutions in a city full of churches and monasteries unless the architect directing the works in the apse, namely Del

Duca, was the same architect who in those same years was working in the building yard for the façade of S. Agata in Alì. Obviously, stylistic aspects also suggest this attribution: i) the trenchant drama of the layout, with the thickening of the angular pilasters (like in the apse of the Maltese Priory); ii) the strong overhang of the cornices, the small windows "hung" to the cornices over the two lateral doors, in analogy with the memorial tablets in the façade of the Apse in Messina but even with the windows of the Apse of San Pietro in Rome; iii) the tendency to keep the two-colour aspect even in the absence of the brick veneer – there were no quarries of clay around Alì – by using two different types of stones: pink limestone for the wall and grey "colombina" (another limestone) for the rest; iv) the "stressed" volutes placed by the two sides of the second order – they are a probable invention by Del Duca;[12] v) the "cubist" brackets over the pilasters, again in the second order[13] (see fig. 16) – some decorations in the Apse of the Maltese Priory and in Porta Pia have the same "cubist" tendency, the swags for example.

Fig. 16. Brackets on the pilasters of S. Agata, Alì

However, the most interesting characteristic of the façade of the church in Alì is its subdivision in three levels. Such a choice is not coincidental, considering that the proportional relationship between the first and the second order is in the ratio 8:5. This ratio is far removed from the tradition of the late Renaissance, nor does it appear during the Roman Baroque, where the 9:8 ratio was preferred. This squatness of the second storey, which, in passing, we meet again in the façades of R. Gagliardi, made it possible to insert the third storey. To recount the historical chronicles correctly, it is necessary to mention that the last order collapsed in the 1783 earthquake and was reconstructed after 1855. Some scholars claim that the reconstruction was very close to the original [Pergolizzi 1971]. However, there are no documents showing the original aspect of the church, the only exception being a painting by Letterio Subba – found in the same church – where the damaged front is visible, but the drawing is so imprecise even in the undamaged parts that it cannot be considered a reliable document. Indications about the existence of the subdivision in three levels arise from the façade of the church of S. Maria delle Stelle (1722-1741) in Militello in Val di Catania.

Fig. 17. Façade of S. Maria delle Stelle in Militello in Val di Catania

Fig. 18. The brackets on the pilasters of S. Maria delle Stelle

Notwithstanding its Baroque exuberance, S. Maria delle Stelle displays in its façade many analogies with the one of S. Agata in Alì: i) the rhythm of the pilasters (fig. 17); ii) the proportional system of the design; iii) the tendency of the central window at the second story to "break" the attic basement – such a breaking takes place in the church of S. Agata in Alì;[14] iv) above all, the strange "brackets" located just below the capital of the pilasters of the second level, like those mentioned previously (fig. 18). These brackets are not present in the other churches of the Sicilian Baroque. Only Del Duca was free enough transfer them from the models of windows by Michelangelo to the primary partition of a façade. The unknown architect of S. Maria delle Stelle" appears to have considered the church in Alì as a model. In actual fact, Gagliardi's drawings for the façade of another S. Maria delle Stelle, the one in Comiso (see fig. 19), also display strong analogies with Del Duca's experiment. But, here, the projecting central bay, from the portal to the belfry, is quite new. Other references should be considered: for example, the church of the Santi Vincenzo ed Anastasio in Rome and the façade of the Cathedral of Enna. However, all these works are later than that of Del Duca. It is also interesting that such a version of Santa Maria delle Stelle in Comiso exists only in Gagliardi's drawing. The actual façade is so far from the original idea that, according to Stephen Tobriner, it isn't a work of Gagliardi at all [Tobriner 1996: 148].

Fig. 19. Gagliardi's drawing of the church of S. Maria delle Stelle in Comiso [Tobriner 1996]

There is a peculiar aspect in Del Duca's *formativity* which is very simple but completely original: the sequence "second order–pediment" is substituted by the alternative sequence "second order–third order–pediment", which is an extension of Cesariano's hypothesis. This substitution and the aim of maintaining a harmonic proportional ratio require, as we saw earlier, a reduction of the height of the second order so that the third order can be inserted without appearing to be a simple addition. The choice of a square proportional grid links Del Duca's experiment to the Renaissance and is also the reason for the strong diminution in the 8:5 ratio.

Fig. 20. Hypothetical guidelines for the façade of S. Agata in Alì. Drawing by the author

Fig. 21. Hypothetical guidelines for Santissima Annunziata dei Teatini. Drawing by the author

In Guarini's Santissima Annunziata the proportional grid is hexagonal and recalls Borromini's experiments. Such a choice makes possible the ratio 7:5, which will become canonical. However, analogies with the façade invented by Del Duca are clear (figs. 20, 21): i) the use of the last order to receive the statue of the saint to whom the church is dedicated; ii) the large window that, together with the decoration below, interrupts the attic basement of the second order, but this feature is present only in the drawing published in 1737; iii) the organization in three quite distinct and proportioned levels, as mentioned repeatedly.

The absence of documents makes it arduous to go further for the moment. However, I believe that the remarks discussed above provide indications for a deeper investigation into the influence exerted by Giacomo Del Duca – an architect with many ideas, although he was not among the highest masters – on the genesis of Baroque architecture, and not only in Sicily.

Notes

1. Gagliardi's first dated drawing of a triple-storey façade dates back to 1744 (S.Giorgio in Ragusa), seven years after the publication of Guarini's *Architettura civile* [1737].
2. The façade of Santissima Annunziata didn't appear in *Dissegni d'architettura civile...* [1686] but only in the posthumous *Architettura civile* [1737].
3. The Jesuits' ability to condition urban planning to their own advantage is well documented; cf. [Aricò & Basile 1998].
4. Actually, by carefully observing the organization of the third level it possible to imagine Guarini's intention to recreate the structure of a lantern projected over a concave surface: the dome is suggested only by the broken tympanum which rises from the attic at the third level. Guarini could refer to many examples (Brunelleschi's lantern in S.Maria del Fiore, the lanterns by Borromini), but it is a work by Giacomo Del Duca, namely the quasi-Baroque lantern in S. Maria di Loreto (see fig. below) which seems to be the closest model. This church will be discussed further on in this paper.

Axonometric projection of the lantern of S. Maria di Loreto by Giacomo Del Duca. Drawing by the author

5. See, for example, the vivid sketch given by Cellini in his *Vita* (vol. II, p. 92): "...*eglino furno*

tanto arditi che e' mi richiesono all'andare in Sicilia, e che mi farebbono un tal patto che io mi contenterei, e mi dissono come frate Giovanagnolo de' Servi [Giovannangelo Montorsoli, Servita] *aveva fatto loro una fontana, piena ed adorna di molte figure, ma che le non erano di quella eccellenzia ch'ei vedevano in Perseo, e che e' l'avevano fatto ricco."*

6. Consider, for example, how much Borromini's style owes to the "ad triangulum" structure described by Cesariano.
7. The windows at the ground floor of Porta Pia have been suggested by them.
8. The rediscovery of winged head of the cherub is due to Del Duca, probably under the influence of his uncle, a priest, the author of a treatise about the angelic hierarchies. This priest was a friend of Michelangelo and pushed hard for the transformation of part of Diocleziano's Thermae into the church of S. Maria degli Angeli, built to a design by Michelangelo.
9. See the discussion of the Medici's coat of arms at Porta Pia in [Benedetti 1973: 52-53].
10. Del Duca was the first architect to design the project for the Jesuits' "Collegio laico".
11. The first attribution dates back to 1894 and is due to the painter Adolfo Romano. Recently, Francesca Paolino suggested, with due caution, that Del Duca was influential in the project of the external aspect of the Mother Church in Alì; cf. [Paolino 1995].
12. This stylistic signature can also be found in the decoration of the courtyard of the Palazzo Piceni in Rome and in the fountain of Neptune in "Villa Mattei", also in Rome.
13. One may conjecture that they were suggested to Del Duca by some drawings of windows by Michelangelo.
14. Del Duca had already used this unorthodox caesura on the interior cornice in S. Maria in Trivio.

Bibliography

ACKERMAN, James S. 1961. *The Architecture of Michelangelo*. London: Harmondsworth.

ARICÒ, Nicola & Fabio BASILE. 1998. L'insediamento della Compagnia di Gesù a Messina dal 1547 all'espulsione tanucciana. *Annali di Storia delle Università italiane* 2. Bologna.

BAGLIONE, G. 1649. *Le vite dei pittori, scultori ed architetti* ... Rome.

BASILE, Francesco. 1960. *Lineamenti della storia artistica di Messina; citta dell'Ottocento*. Messina: Leonardo

BENEDETTI, Sandro. 1973. *Giacomo Del Duca e l'architettura del Cinquecento*. Roma: Officina.

BLUNT, Anthony. 1968. *Sicilian Baroque*. London: Weidenfeld & Nicolson.

CESARINANO, Cesare. 1521. *Vitruvio De architectura*. Como.

DEL DUCA, Giacomo. 2004. *L' arte dell'edificare*. Francesca Paolino, ed. Introduction by Sandro Benedetti. Messina: Società Storia Patria Messina.

GUARINI, Guarino. 1686. "*Dissegni d'architettura civile, et ecclesiastica, inventati, e delineati dal padre d. Guarino Guarini*. Torino: Per gl'Eredi Gianelli.

———. 1737. *Architettura civile*. Torino, G. Mairesse all'insegna di Santa Teresa di Gesú.

MEEK, H. A. 1988. *Guarino Guarini and his architecture*. New Haven: Yale University Press.

PAOLINO, Francesca. 1995. *Architetture religiose a Messina e nel suo territorio fra Controriforma e tardorinascimento*. Messina: Società Storia Patria Messina.

PERGOLIZZI, Fortunato. 1971. *Architettura michelangiolesca in Sicilia : la Chiesa di Alì*. Messina: Società Storia Patria Messina.

TOBRINER, Stephen. 1996. Rosario Gagliardi and the Development of the Sicilian Tower Façade. Pp.141-155 in *Rosario Gagliardi e l'architettura barocca in Italia e in Europa*. Lucia Trigilia, ed. *Annali del Barocco in Sicilia* 3.

About the author

Pietro Totaro received a master's degree in civil engineering from the University of Rome "La Sapienza" and then a Ph.D. in "Ingegneria Edile: Architettura e costruzione" (2001) from the University of Rome "Tor Vergata". From 1996 to 2002 he was assistant, first in the courses of "Composizione architettonica" at the University of Rome "La Sapienza", then in the course of "Architettura Tecnica I" at the University of Messina. He currently works as a professional in Messina, continuing his studies on Baroque architecture.

Ntovros Vasileios

Dokos-Chalkida
34100 Chalkida
GREECE
dovarch@gmail.com

Keywords: Guarino
Guarini, Baroque
architecture, Chiesa di San
Lorenzo, Gilles Deleuze,
fold,

Research

Unfolding San Lorenzo

Abstract. This paper proposes a "reading" of the church of San Lorenzo in Turin, designed by Guarino Guarini, through the philosophical notion of "fold" introduced by Gilles Deleuze. The paper consists of two parts. The first part contains an exploration of the notion of "fold" in architecture and in philosophy and examines the use of the fold in the theory of Baroque architecture as well as the range of this new tool in architectural practise in contemporary architecture and in philosophy and examines the use of the fold as fundamental condition for understanding Baroque era. The second part contains the application of the notion of fold as a philosophical and conceptual framework for the "reading" of the chapel.

Fold as conceptual framework

The contemporary architecture of the Fold

Towards the close of the twentieth century a number of architects believed that the discussion about architecture should be enriched with new philosophical tools. A group of prominent architects – G. Lynn, K. Powell, P. Eisenman, J. Kipnis, I. Rajchman, B. Shirdel, F. Gehry – created an informal manifesto, the May 1993 issue of *Architectural Design*, entitled "Folding in Architecture" (fig. 1). The French philosopher Gilles Deleuze was represented in this manifesto through the first chapter of his book *Le Pli, Leibniz et le Baroque*, translated in English as *The Fold, Leibniz and the Baroque* [1993].

Fig. 1. The cover of *Architectural Design*, "Folding in Architecture"

Editor Greg Lynn clearly states the effort to detect new techniques and strategies for the confrontation of complexity: "For the first time perhaps, complexity might be aligned with neither unity nor contradiction but with smooth, pliant mixture." Folding is perhaps presented in architecture as a different way, "to integrate unrelated elements within a new continuous mixture" [Lynn 1993: 8].

Nexus Network Journal 11 (2009) 471–488
DOI 10.1007/s00004-009-0008-6; *published online* 5 November 2009

In Post-Modernism and Deconstructivism, complexity and contradiction emerge through conflict. On the contrary, contemporary attempts aim to fold locations, materials and programs into architecture via the incorporation of exterior forces "while maintaining their individual identity" [Lynn 1993: 10]. It is significant to note here that there is a risk of misunderstanding the notion of fold as simply a new morphological tool and of confusing its use with its literal significance as mere folded figures. Greg Lynn seeks to understand curvilinear logic, contrary to the curvy style of a new Baroque expressionism. Furthermore, he believes that the appropriate geometry for the bent and twisted forms is the field of topology, which has the ability to organise different elements into continuous spaces.

Thus the meaning of Fold – the act of folding – can be summarized in terms of a spatial tool used by contemporary architecture and originating from Deleuze's philosophical work: "Folding" constitutes a tool for confronting architectural problems. It recommends a new strategy to manage difference and complexity. The final result is unique and is composed of a number of dissimilar elements in a continuous relation in which they maintain their particular characteristics. Through the use of topology and curvilinear logic, "Folding" tries to incorporate all the design parameters of architectural practice into a process of multiple connections – folds – up to the production of a final proposal.

The philosophical origins of the Fold

The book *Le Pli: Leibniz et le Baroque* was published in 1988 in France, while its translation as *The Fold: Leibniz and the Baroque* was realised in 1993, the same year as the issue "Folding in Architecture". It has been called the most "personal and authentic" [Conley 1993: xi] text of Deleuze and is part of his wider philosophical approach.[1] Moreover, while on one hand the content of the book speaks for a tool – a way of regarding philosophically the things – on the other hand, the book itself is characterized by this. In other words, at the same time that Deleuze explains what the fold is, he folds the book's contents.

Initially the book *The Fold* can be considered as a particularly clear-sighted contemporary critique of the Baroque era. By using and analyzing music, mathematics, science, painting, theatre and costume, Deleuze makes unique observations about the period and its cultural identity. There have been attempts to interpret the architecture of the seventeenth century in later periods, resulting in the gradual recognition of Baroque as a distinct period at the beginning of the twentieth century. However, Deleuze neither focuses on a concrete period, nor reduces Baroque to an architectural style or to a limited geographical area. Instead, he declares, "The Baroque never existed" [Deleuze 1993: 33], going on to say that if it has a certain reason to exist, this should be given via an idea, a concept. Thus, from Deleuze's point of view, "the criterion or operative concept of the Baroque is the Fold, everything that includes, and all its extensiveness" [Deleuze 1993: 33].

How Deleuze is led to the Fold as the essential criterion of Baroque

Deleuze is seeking the beginnings and the philosophical framework in which Baroque is inscribed. He maintains that, more than any of the other philosophers, mathematician and philosopher Gottfried Wilhelm von Leibniz (1646-1716) epitomizes in his works the real content of his era. Deleuze finds in Leibniz a Baroque regime "where things can be continuous even though they are distinct and where what is clear or clarified is only a region within a larger obscurity" [Rajchman 1993: 62]. This approach is characterized as the antithesis of Cartesian philosophy, recommending a new possibility in which the things can be continuous and distinguishable instead of being clear and distinguishable. Thus, it recognizes multiplicity as the co-presence of differences and not as the sum of parts.[2]

Deleuze generally distills the contribution of the Baroque in art and the contribution of Leibnizianism in philosophy to six points. These are:[3]

1. The fold: the infinite work or process. The question here is not how to complete it, but how to continue it, how to bring it to infinity. It determines and materializes Form; it produces a form of expression, the curve with the unique variable.

2. The inside and the outside: the infinite fold separates or moves between matter and soul, the façade and the closed room, the inside and the outside. Because it is a virtuality that never stops dividing itself, the line of inflection is actualized in soul but realized in matter, each one on its own side. Baroque architecture is forever confronting two principles, a bearing principle and a covering principle.

3. The high and the low: the perfect accord, or the resolution of tension, is achieved through the division of the world into two levels, the two floors being of one and the same world. The façade-matter goes down, while the soul-room goes up above. Pleats of matter exist in a condition of exteriority, folds of the soul are found in a condition of enclosure.

4. The unfold: this is not the opposite of the fold, but the continuation or the extension of its act. It is the result of the infinite work of the fold.

5. Textures: the way through which a material is folded is what constitutes its texture. Everything is folded in its own manner. Folds of matter and textures become expressive when related to several factors.

6. The paradigm: the search for a model of the fold is directly related to the choice of a material. The composite materials of the fold should not conceal the formal element or form of expression. The paradigm becomes "mannerist" and proceeds to a formal deduction of the fold.

Fold, to fold, in the work of Deleuze and as a tool of analysis

The Fold is relation, either as a term or as a strategy [Benjamin 1997]. It creates connections among differentiated elements comprising a unified complex totality. Characteristics of the fold are *com*-plication, *im*-plication and all the differentiations of the relations between the One and the Multiple. Moreover, the multiple is not only what has many parts but also what is folded in many ways [Deleuze 1993: 3]. The effective space is always the in-between place among the elements, where limits can become imperceptible and differential relations can be activated. Inside the fold all the things are continuous and distinct. The Unfold is the extension of the act of Folding and not its opposite. Folding-Unfolding does not simply mean tension-release, but also enveloping-developing or involution-evolution.

My point of view starts from a philosophical and conceptual framework of the fold and not from architectural-spatial practise, so that it can be connected with the spatial qualities of the church of San Lorenzo.

"Reading" San Lorenzo

Presence of elements: the folds

The chapel of San Lorenzo constitutes an architectural organism.[4] It is composed of many different elements found either in their simplest form, in fundamental combinations, or in complicated relations. The intention is to recognize the first complex folds, which are combinations of heterogeneous elements in multifunctional parts. In architectural terms, the aim is to locate the combinations of architectural parts that comprise different styles and orders.

Renaissance Folds. The arrival of Guarini in Turin in 1666 is accompanied by a radical change of the existing ground plan of the church, which was based on a Latin cross. As a new organisational diagram Guarini introduces the centripetal shape of a cube and a semi-spherical dome which appears to be supported internally by big piers (fig. 2a). Such a diagram emanates from the architectural discussions of the fourteenth and fifteenth centuries about the Ideal and the Harmony in the planning of temples. More consistent with the heritage of the Renaissance, Passanti [1990] believes that the temple is organised around the form of a Greek cross in its base (fig. 2b). It is significant to note here that during the Renaissance, more temples were built in the shape of polygons or Greek crosses than with an ideal circular form [Wittkower 1998: 39]. Finally, despite the fact that Guarini is thoroughly familiar with all the orders and the ideal proportions of Renaissance, he chooses to adopt the earlier traditions to suit his own purposes in San Lorenzo.[5]

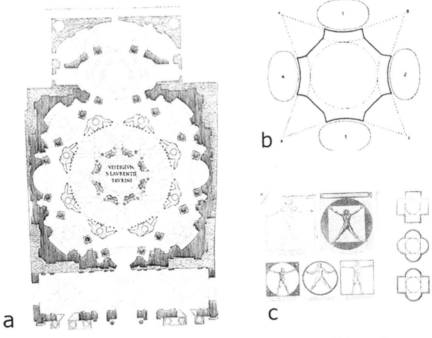

Fig. 2. Renaissance folds. a) plan of the chapel San Lorenzo; b) Passanti's explanation diagram; c) ideal Renaissance plans.

Gothic folds. The first of his times to do so, Guarini refers to Gothic architecture with a particularly appreciative critical glance and admiration. Comparing the objectives of Romanesque and Gothic architecture, he underlines that the objective of the Gothic style is to construct very robust buildings that appear fragile: "(they) appear feeble and the fact that these stand, looks like a marvel" [Guarini 1968: 209]. He applies this strategy in San Lorenzo. It is a source of real wonder that the forces of the dome are balanced by small elements, while the asymmetrical proportion of height between them creates more concern, pressure, imbalance and sense of particular boldness (fig. 3a).

The use of slender stone arches closely links Guarini to the Gothic style. His travels in France during the period 1662-1665 gave him the opportunity for contact with not only many Gothic temples but also with the work of Desargues and Derand in stereotomy. Guarini dedicates one-third of his treatise to sketches on the new science, describing methods for calculating volumes, and designing and cutting solids, evolving the know-how of his era in a way that was different from that of the more practical Borromini and the more artistic Bernini (fig. 3b). Thus, stereotomy made it possible to reintroduce and study Gothic techniques during the French Gothic rebirth of the seventeenth century and in Guarini's architecture, even though the necessary "men of genius" – craftsmen with their unique mastery to create any form of arc, dome, volumes mainly from stone – no longer existed [Wittkower 1974: 96].

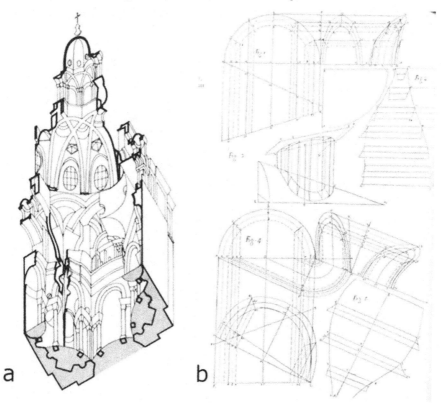

a   b

Fig. 3. Gothic folds. a) Axonometric of the chapel; b) Drawings on stereotomy from Guarini's *Architettura civile*

Islamic folds. Giedion claims that "the dome of San Lorenzo would have never been conceived, if Guarini had not seen the domes of the Mihrab of the mosque Al Hakem in Cordova" [Giedion 1967: 126] (fig. 4). Although the similarities – the techniques of manufacture, the registration in a square base, the system of interlaced arches, overhead lanterns – are important, Meek [1988], Wittkower [1965], and Gasparini and Volpato [2003] believe that the differences are even more important. Therefore the big transparent dome of Guarini cannot be associated with that of the mosque. We are not certain if Guarini saw ever the dome in Cordova, but it is known that he travelled in Andalusia and in Castile, and that he also came into contact with Islamic architecture during his stays in Naples, Sicily and Messina. This might explain his adaptation of the star symbol, a formal Islamic product, in many expressions and on many levels in his work.[6] In San Lorenzo, eight stars can be found: four star hexagons in the small domes of the chapels, two star octagons in the big domes, one star hexagon in the dome of the presbytery and a star with sixteen points in the floor.

Fig. 4. Islamic folds. Comparison between the dome of San Lorenzo (left) and the Al Hakem mihrab in Cordova (right)

Partial folds. Guarini uses also other individual folds of smaller scale. Consequently we can observe the introduction of the Corinthian order in columns and the corresponding entablatures, the attic storeys, the campanile in the form of Dorian column, the Serlianas, typical characteristics of Northern Italian architecture (fig. 5). Furthermore according to Wittkower, the stimulus to conflict on the level of interior decoration ties the church of San Lorenzo to the mannerist tradition [Wittkower 1965: 271]. While the exterior was initially designed with a formal Corinthian rhythm, ultimately a unified design for the facades of buildings in Piazza Castello by Vittozzi was applied.

It seems clear that there is a noticeable richness of elements in various combinations. Their presence in San Lorenzo constitutes a particular architectural situation, because while together they are completed wholes, they are also autonomous. In their interior they contain individual distinguishable elements, where they recommend the identity of a particular composition in certain historical periods. Their use in the church, aside from Guarini's ability and his knowledge of architectural history, also reveals the unique character of the Baroque. As Deleuze says:

> The Baroque refers not to an essence but rather to an operative function, to a trait. It endlessly produces folds. It does not invent things: there are all

kinds of folds coming from the East, Greek, Roman, Romanesque, Gothic, Classical folds. ... Yet the Baroque differentiates its folds, pushing them to infinity, fold over fold, one upon the other [Deleuze 1993: 3].

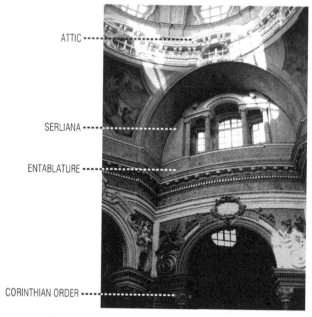

ATTIC

SERLIANA

ENTABLATURE

CORINTHIAN ORDER

Fig. 5. Partial folds. Photograph by the author

Relation of elements-Differential relations: The mathematics of folding

From the analysis of the elements, a question emerges: How do the disparate elements, the infinite folds, imply the total, the entirety of the chapel? In other words, how can the individual fragmentary perceptions/sensations that are constituted by every kind of fold, lead to a clear perception of the chapel?

Leibniz specifies that the "relation of inconspicuous perceptions to conscious perception does not go from part to whole, but from the ordinary to what is noticeable and remarkable" [Deleuze 1993: 87]. Deleuze goes on to say that the conscious perception is produced when at least two heterogeneous parts enter into a differential relation that determines a singularity. Expressed in mathematic symbols, that is dx/dy. For example the yellow and the blue that determines the green,

$$d \text{ yellow } / d \text{ blue} = \text{green [Deleuze 1993: 88]}.$$

This is the basic idea of what we call differential calculus.[7] Together, differential calculus and integral calculus comprise the mathematical field of infinitesimal calculus or, more simply, calculus. The calculus was discovered simultaneously by Leibniz and Newton from completely different starting points[8] and it is a genuine discovery of seventeenth century which expresses concern for infinity. For the first time in the history of human thought, the idea of infinite subdivision was adopted in tiny elements that give direct results either through differentiation (differential calculus) or through integration (integral calculus), with a plethora of applications in geometry, in mathematics, in physics. Knowledge of the microscopic shows the way for conscious perceptions, in other words the "macroscopic form always depends on microscopic processes" [Deleuze 1993: 88].

Coming back to San Lorenzo, it is now possible to comprehend how Guarini accomplishes the unity of all the different elements. For example, one of the bigger problems that he faces is how an octagonal ground plan is connected without discontinuities to the form of the Greek cross and then to the elaborate domes. The architectural treatment, particularly complicated but also successful, is focused in the four corners of the temple, in the four big piers. Taking three sections at different heights, a completely different picture of shapes appears (fig. 6). There are not precise points where the passage from the one element – form – into the other takes place, but there is a continuous differential relation between them. "Differential relations intervene only in order to extract a clear perception from minute, obscure perceptions... an unconscious psychic mechanism that engenders the perceived in consciousness" [Deleuze 1993: 95-96].

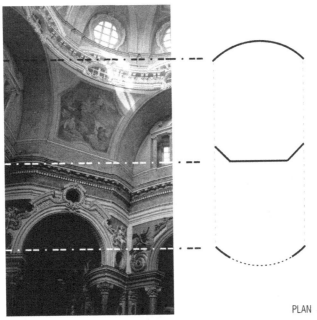

PLAN

Fig. 6. Diagrammatic sections at different heights. Photograph by the author

In differential relations, the "remarkable" comprises a singularity, but not the only one in the Baroque kingdom.[9] The point of inflection is also an intrinsic singularity, "it corresponds to what Leibniz calls an ambiguous sign" and what Klee calls "a site of cosmogenesis, a no dimensional point, between dimensions" [Deleuze 1993: 15]. In this point the curve turns from concavity to convexity and vice versa.

In the ground plan of San Lorenzo, a sinuous line runs along the perimeter walls at roughly eye level. In this continuous perimeter we can identify eight unique turning points where the convexity changes to concavity and where certain characteristics can be assembled (fig. 7).

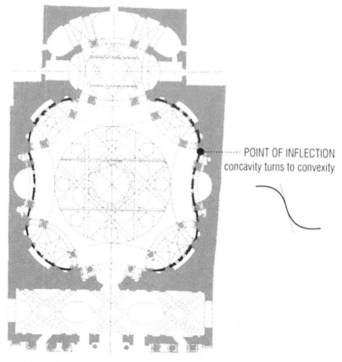

POINT OF INFLECTION
concavity turns to convexity

Fig. 7. One of the eight points of inflection that can be identified in the plan of San Lorenzo

spring the vaulted ribs

the intersection point
of the hidden arches

angles in the
continuous serliana

the four angular piers
are originated/eliminated

Fig. 8. The characteristics of the points of inflection

In the projected vertical axis starting from these points (fig. 8):

- the four angular piers are originated or eliminated;
- oblique angles appear from the continuous serliana, thus a totally different geometry from all the other smooth and curvilinear architectural gestures;
- the big arches of the dome spring and the walls of the main branches of the Greek cross begin;
- we find the intersection points of hidden arches that support the dome (see the next section, "Pleats in matter").

However, these points can be validated not only as "pure events of the curve" [Cache 1995: 17], but mainly as "point-fold" [Deleuze 1993: 14], where heterogeneous and distinct elements are presented as united and continuous.

Materiality: Pleats in matter

San Lorenzo is characterized by a particular wealth in the use of materials, characteristic of the entire Baroque period. We find white, black, red marbles in various places, stucco, gilt decorative patterns, glass, slabs of local stone. To each of these materials is attributed one final texture which reveals what they are made of and how they are worked. "A general rule for the way a material is folded is what constitutes its texture...everything is folded in its own manner" [Deleuze 1993: 36]. Observing the interior of the temple, the Palladian patterns, the windows, the decorations and the dome, the spectator gets the impression that the materiality becomes a way of managing all the aesthetic and functional gestures, a way of expressing a wider folding.[10] However, the spectator can also see that the interior of the temple, up to the height of the dome, serves no static purpose. At the points where one would reasonably expect the dome to be supported, Guarini places oval openings – in the keys of the small arches, in the centres of the pendentives, in the keys of the four big arches – undermining any conventional static function (fig. 9). Thus we must look elsewhere for the structural organism, which is concealed. Guarini applies a complicated static system inspired by his knowledge of Gothic architecture. The actual support at the second level[11] is based on four huge masonry arches that spring from the four corners of the square ground plan. Four smaller diagonal masonry arches originate at the third points of the biggest arches, so that they are connected with the main arches at eight points. From these eight points, four pairs of interlaced vaulted ribs constitute the real static structure of the dome (fig. 10).

Taking the above into consideration, we realize that the external surfaces are simple in comparison to the complicated internal static mechanism. The relation that is developed between them is very powerful. The external construction, aside from the rigidity that it adds to the internal structure, indicates the concealed static system through the decorative elements, like the attics, while preparing its revelation at the superior level. In reverse, the concealed mechanism, by undertaking all the static forces, leaves the exterior surface free to move upwards to create and structure all the perceptual sensations of the internal. Thus "the two kinds of force, two kinds of folds – masses and organisms – are strictly coextensive" [Deleuze 1993: 9], "the visible and the legible, the outside and the inside, the façade and the chamber are, however not two worlds, since the visible can be read and the legible has its theater" [Deleuze 1993: 31].

OPENINGS

Fig. 9. Openings instead of structural solidity. Photograph by the author

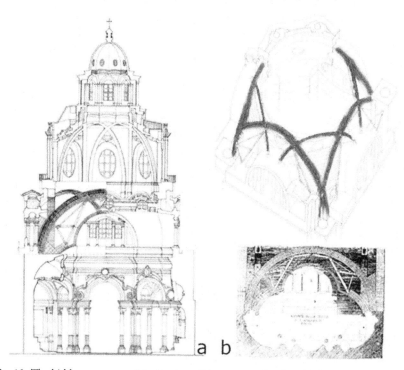

a b

Fig. 10. The hidden structure. a) Indication of the structural arches; b) the engraving of 1677
entitled "Key to the Dome of San Lorenzo In Turin"

However the connection in the same world does not become static and none of the pieces loses any of its substance. The particular characteristics are inherently present at their union:

> ...the infinite fold separate or moves between matter and soul, the facade and the closed room, the inside and the outside... Up to now Baroque architecture is forever confronting two principles, a bearing principle and a covering principle. Conciliation of the two will never be direct, but necessarily harmonic, inspiring a new harmony [Deleuze 1993: 35].

Light and depth: Folds of the soul

For Leibniz the soul and the body are different expressions of the same substance, depending on their degree of development. In contrast, in almost the same period, the philosopher Descartes accepts an absolute dualism between body-soul that is condensed in extent-intellect. In the Baroque period, the material-primal force duality is what replaces the Cartesian matter-form duality, and constitutes a different expression of body-soul.[12] In architectural terms, and more specifically in the period that we are examining, light is considered as a primal force. Its management is a particular factor that invigorates matter. It does not have concrete form; the form is given and its shape depends on the built environment. Looking at San Lorenzo, Guarini seems to know very well that light and depth can connect or differentiate spaces.

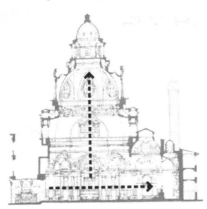

Fig. 11 (left). The main axes of manipulation of light and depth
Fig. 12 (right). Interior of the chapel looking towards the retrochoir

Two axes – horizontal and vertical – can be observed in the use of light and the manipulation of depth (fig. 11). Along the horizontal axis, starting from the entrance of the chapel, the light disappears by degrees into the dark sanctuary and then into the darker retrochoir (fig. 12). This is the idea of *chiaroscuro*,[13] a Baroque characteristic: "the white is progressively shaded giving way to obscurity... clarity endlessly plunges into obscurity" [Deleuze 1993: 32]. In reverse, along the vertical axis the light is successively increased by degrees as it goes up: from the small chapels in shadows, to the bright main space, and finally to the translucent domes. Furthermore, by means of several architectural mannerisms – such as the evacuation of the mass from the piers and the use of white statues in front of walls with black marbles – Guarini gives the temple a sense of depth. This is achieved along both axes, since the spectator either goes down to the regime of darkness or opens up to the regime of light (fig. 8).

We can summarize by saying that the intensity of light in relation with the height and the depth of the chapel presents multiple differentiations. However, it's not possible to define variant categories or to distinguish clearly the different qualities which are achieved. Light is found in a status of continuous folding with its internal ingredients according to the chapel's parts, according to the pleats of matter.

The "two floors": a Baroque allegory

Deleuze schematizes in a diagram the main relations that he reads in the Baroque situation. Thus the "Baroque house" is comprised of two main floors: the lower has some openings and communicates with the environment, while the upper is interiorized (fig. 13). Matter, organic and inorganic, is located in the lower level and the soul in the upper level.[14] The pleats of matter as an infinite room for reception and receptivity, and the folds of soul as closed interiority, are found in two different levels of the same world, of the same house, as matter and soul don't have an exact point of division. For Leibniz the two "floors" are distinct but also inseparable through the presence of the upper in the lower. "The upper floor is folded over the lower floor. One is not acting upon the other, but one belongs to other, in a sense of a double belonging" [Deleuze 1993: 119].

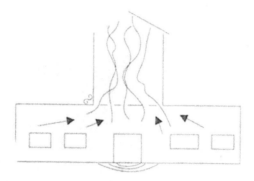

Fig. 13. Diagram of the Baroque house, an allegory of Deleuze

The diagram that is adopted from Deleuze reveals an interpretation that would not symbolize the idea of Baroque, but would mainly act as an allegory. He borrows from Walter Benjamin the particular distinction that the symbolic combines the eternal and the momentary while the allegoric uncovers nature and history according to the order of time. Reading Walter Benjamin, Deleuze underlines his idea that in Baroque "…allegory was not a failed symbol or an abstract personification but a power of figuration entirely different from that of symbol" [Deleuze 1993: 125]. Contrary to symbolism, which tries to conceive the idea of the object through abstraction, allegory seeks to understanding the idea of the object through the network of its multiple connections with its environment. Deleuze supports the second field and he declares with emphasis that "instead of sticking to abstractions, we have to restore the series" [Deleuze 1993: 56].[15]

Coming back to San Lorenzo, an abstract schema is not difficult to form. The chapel is divided into two levels, since the confused surface stops at mid-height to give priority to the true structure of the perforated domes. Meek recognizes an opposition between "the pressures and the deceit of the lowest zone, and the stripped clarity and the truth of the celestial regions, which may in itself have theological or rhetorical implications" [Meek 1988: 60]. Infinite God can be identified with the infinite sense of continuous domes, the truth, the clarity, the light, the miracle, while the believer and the surrounding earth are found in the regime of darkness, deceit and illusion. The only way to escape is upwards, to the salvation of the soul (fig. 14). Most Guarini scholars agree that the chapel conceives and embodies the notion of infinity, the typifying characteristic of God, in a unique architectural manner. Thus the chapel effectuates, not only functionally but also notionally, the Catholic Church's message of Salvation.

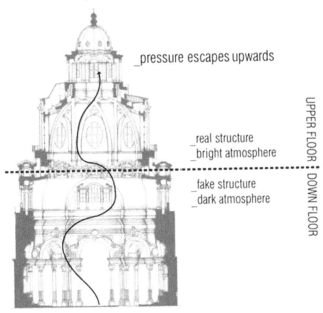

Fig. 14. Abstract symbolic scheme

Contrary to the above abstract schema, let us restore the series of things as they have been described up to this point.

The church of San Lorenzo is divided vertically into four equal parts:[16] the chapels, the Greek cross, the big dome and the two smaller domes. In the two lower parts the superficial construction is internally folded in a unique manner. In the two upper parts the real structure is revealed and combined with the continuous penetration of light, creating another unified ensemble. We can identify the two primary unities with the main principle of their internal folds, matter and soul.

The pressure of the lower unity is as much the result of the twisted surfaces as it is of the darkness. The pressure escapes upwards but not exactly to the insubstantial peak of the church, not to infinity. The attention of the spectator is concentrated on the marvelous dome because of its structure and its astonishing geometry. The sixteen woven vaulting ribs attract the spectator's gaze. If the spectator tries to follow a rib of the dome to its springer, he will meet the starting point of the twin arch; following again the trajectory of the arch until it comes down on another springer, "he will find himself progressively re-launched from one springer to other, without ever coming to the end of his course" [Meek 1988: 48]. He is trapped in a seemingly infinite weave, a complicated continuum with no beginning and no end. The tension is balanced when the gaze manages to return from a springer to the ring of the dome and then to the schema of the Greek cross, thus when it stays *in the middle of chapel.*

In this region, between the second and the third part of the chapel, between the transition to the dome and the dome itself, a continuous movement-encirclement of gaze and senses can be located. In this region the celestial light from the internal domes is diffused downwards to more public spaces, while the matter is filtered and eliminated upwards to more private regions. Matter and rays of light are interlaced in such an admirable way that they interchange their attributes: the domes are dematerialized, their dimensions are opened to the sky, while the light conceives dimensions and becomes architecture. "Is it not in this zone ... that the upper is folded over the lower, such that we can no longer tell where one ends and the other begins, or where the sensible ends and the intelligible begins?" [Deleuze 1993: 119] (fig. 15).[17]

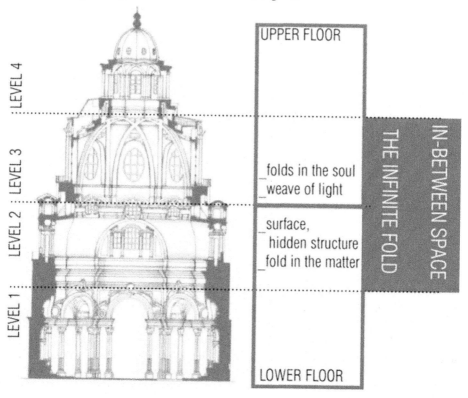

Fig. 15. Allegoric scheme, a different reading of the chapel

Thus in contrast with the previous symbolic scheme, we try here to recommend an allegory. The striving to express infinity – the meaning and typifying characteristic of the God – is no longer the main factor that determines the architectural process. The idea is found in the infinite fold, in the infinite creation of continuous connections, in the infinite negotiation of a solid relation among disparate elements. The idea is not found in the contrast of the upper level-lower level, light-dark, true-fake, infinite-finite, but rather in the manipulation of contradictions through their unity in difference. The idea is not found in the static symbolism of the meaning, but in the dynamic presence of the relation. Guarino Guarini manages, as he does in all the other fields of his multifaceted activity, to ascribe the dynamic and indefinite together with specific forms in a single continuous relation, in a single spatial proposal.

Acknowledgments

The article "Unfolding San Lorenzo" is a summary of my presentation in the Symposium "Guarino Guarini's Chapel of the Holy Shroud in Turin: Open Questions, Possible Solutions", 18-19 September 2006, Turin, Italy. The presentation was based on my pregraduate research thesis "Unfolding San Lorenzo" at the School of Architecture of Aristotle University of Thessalonica, under the supervision of Ass. Professor Vana Tentokali.

Notes

1. The work of Deleuze is conditioned by a particularly developed relation with the history of philosophy. Very often the main content of his writings is the revaluation of previous philosophers and the actualisation of them via the exertion of new ideas and the stir of old ones; cf. [Roffe 2002].
2. Andrew Benjamin [1997] points out the Cartesian perception of the complex as the result of the simple instead of Leibniz's position where the emphasis is located in the co-presence of the finite and infinite.
3. See the six points of Deleuze's detailed description [Deleuze 1993: 34].
4. Guarini is affronting the buildings as "vibrated organisms that are open and indeterminate but they are characterized by specific morphological principles" [Norberg-Schulz 1999: 116]
5. "...but criticizing discreetly... I could correct an old rule and add another one" [Guarini 1968: 15].
6. See H. A. Million's Ph.D. thesis, "Guarino Guarini and the Palazzo Carignano in Turin" [1964], where he attributes a symbolic meaning to Guarini's use of the star in Palazzo Carignano.
7. Differential calculus refers to the calculation of momentary change of the function – F(x) – via the ratio of imperceptibly small changes of the Function – dF(x) – with the corresponding change of the variable – x – when this tends to 0, that is to say, dF (x)/ dx.
8. For many years a controversy raged between the two famous scientists and, more generally, between the continental European and English scientific communities, about who deserved credit for the discovery of the calculus. However, Leibniz arrived at differential calculus via geometrical approaches and as the proof in the microscopic scale of the existence of infinitesimal units. On the other hand, Newton, professor at Cambridge, was led to it via his studies on the mathematic description of the momentary change in the motion of natural bodies.
9. Deleuze mention three kinds of singularities: 1) the point of inflection; 2) the axis of the curve from the concave side insofar as the monad's point of view is determined; 3) what is remarkable, according to differential relations. He underlines that in the deepest Baroque regions, and in the deepest Baroque knowledge of the world, the subordination of true to what is singular and remarkable is being made manifest [Deleuze 1993: 91].
10. The use of different materials is not a problem in the meaning of the fold, because "two parts of really distinct matter can be inseparable ... by the action and pressure of surroundings forces" [Deleuze 1993: 5-6].
11. The earliest knowledge about this second static system came from a copperplate engraving of 1677 entitled "Key to the Dome of San Lorenzo In Turin", discussed by Meek [1988] and by Gasparini and Volpato [2003]. Only recently has extensive work of restoration made it possible to describe the complete static system with certainty. See [Quarello 2009] in this present issue of the *Nexus Network Journal*.
12. According to Deleuze, "in the Baroque the coupling of material-force is what replaces matter and form" [Deleuze 1993: 35].
13. Deleuze's definition of this word is "the way the fold catches illumination and itself varies according to the hour and light of day" [Deleuze 1993: 37].
14. This relation can be recognized in the paintings of Baroque masters like Tintoretto and El Greco. Especially in the *Burial of Count Orgaz* El Greco organizes his amazing composition

around the terrestrial lament and the faith of the celestial host, in this separation of the soul from the body. According to Deleuze, "the Baroque world is organized along two vectors, a deepening toward the bottom, and a trust toward the upper regions" [Deleuze 1993: 29].

15. He means the series of sequentially tiny sensations-perceptions-actions that constitute an event

16. The number of the divided parts differs in the several readings according to the strategy used in the analysis. Thus Robison [1985] and Gasparini and Volpato [2003] refer to four parts, Wittkower [1965] and Giedion [1967] recognize three, while Passanti [1990] sees only two. Meek [1988] tries to avoid the separation, but at the end his analysis contains three parts.

17. In his allegorical schema Deleuze is referring precisely to the relation of the two floors of the intermediate zone.

Bibliography

BENJAMIN, Andrew. 1997. Time Question Fold.
   http://www.basilisk.com/V/virtual_deleuze_fold_112.html.
CACHE, Bernard. 1995. *Earth moves: The furnishing of territories.* Trans. Anne Boyman. Cambridge, MA: MIT Press.
DELEUZE, Gilles. 1993. *The Fold, Leibniz and the Baroque.* Introduction and comments, Tom Conley. Minneapolis: University of Minnesota Press.
ECO, Umberto. 1972. A Componential Analysis of the Architectural Sign/Column/. *Semiotica* **2**. Netherlands: Mouton & Co. N. V., Publishers
EISENMAN, Peter. 1993. Folding in Time. *Architectural Design* (A.D.) **102**: 23-25.
FLEMING, J., H. HONOUR, N. PEVSNER. 1972. *The Penguin Dictionary of Architecture.* Harmondsworth: Penguin.
GASPARINI, Elena and Ramona VOLPATO. 2003. Guarino Guarini e la "Chiave della cupola di San Lorenzo a Torino". La complessità nascosta. *Tecnologos.*
   http://www.tecnologos.it/Articoli/articoli/numero_009/10guarini.asp.
GIEDION, Sigfried. 1967. *Space, Time and Architecture: The Growth of a New Tradition.* Cambridge, MA: Harvard University Press.
GUARINI, Guarino. 1968. *Architettura Civile.* Introduction by Nino Carboneri, comments and appendix by Bianca Tavassi La Greca. Milan: Edizioni il Polifilo.
Guarino Guarini e L'internazionalità del barocco. 1970. Atti del convegno internazionale 1968. 2 vols. Turin: Accademia delle Scienze.
LYNN, Greg, ed. 1993a. *Folding in architecture.* Architectural Design Profile 102. London: Academy Editions.
———. 1993b. Architectural Curvilinearity. The Folded, the Pliant and the Supple. *Architectural Design* (A.D.) **102**: 8-15.
KIPNIS, J. 1993. Towards a new Architecture. *Architectural Design* (A.D.) **102**: 41-49.
MEEK, H.A. 1988. *Guarino Guarini and his Architecture.* New Haven: Yale University Press.
MILLON, Henry. A. 1965. Book review of G. M. Crepaldi, *La Real Chiesa di San Lorenzo in Torino. Art Bulletin* **47**: 531-532.
———. 1999. *Triumph of the Baroque: Architecture in Europe*, 1600-1750. London: Thames & Hudson.
MÜLLER, W. 1968. The Authenticity of Guarini's Stereotomy in his *Architettura Civile. Journal of the Society of Architectural Historians* **26**, 3 (October 1969): 202-208.
NORBERG-SHULTZ, Christian. 1999. The Age of the Late Baroque and Rococo. Pp. 113-131 in *The Triumph of the Baroque. Architecture in Europe 1600-1750,* Henry A. Millon, ed. New York: Rizzoli.
PASSANTI, M. 1990. *Architettura in Piemonte: da Emanuele Filiberto all'Unita d'Italia (1563-1870): genesi e comprensione dell'opera architettonica.* Torino: Allemandi.
QUARELLO, Ugo. 2009. The Unpublished Working Drawings for the Nineteenth-Century Restoration of the Double Structure of the Real Chiesa di San Lorenzo in Torino. *Nexus Network Journal* **11**, 3 (Winter 2009)
RAJCHMAN, J. 1993. Out of the Fold. *Architectural Design* (A.D.) **102**: 61-63.
ROBISON Elwin. C. 1985. Guarino Guarini's Church of San Lorenzo in Turin. Ph.D. thesis, Cornell University.

ROFFE J. 2002. Gilles Deleuze. The Internet Encyclopedia of Philosophy. http://www.iep.utm.edu/d/deleuze.htm#Biography.

WITTKOWER, Rudolf. 1965. *Art and Architecture in Italy, 1600 to 1750*. Harmondsworth: Penguin Books.

———. 1974. *Gothic vs. Classic: Architectural Projects in Seventeenth-Century Italy*. New York: G. Braziller.

———. 1998. *Architectural Principles in the Age of Humanism*. London: Academy Editions.

About the author

Ntovros Vasileios was born in Athens, Greece, in 1981. He studied architecture at the Aristotle University of Thessaloniki (1999-2005) and then proceeded to post-graduate studies for a Master's of Advanced Architecture "Self Sufficient Habitat" at IaaC in Barcelona (2006-2007). Since 2004 he has successfully participated in national and international competitions (2008 second prize, "D. Areopagitou 2008"; 2006 first prize "Celebration of the Cities" UIA with the project "The Intelligent Urban Void"). He has presented several papers at international symposiums and conferences, and has published several papers in Greek and international magazines such as *Architectural issues* and *L'architecture d'aujourd'hui*. Since 2006 he has taught occasional seminars such as EASA summer campus and IaaC's WAW (Weekend Architectural Workshop). He is currently collaborating with the Barcelona architectural firm F451Arquitectura and is working on several projects in Greece.

Book Review

Michael Ostwald

The Architecture of the New Baroque : A Comparative Study of the Historic and the New Baroque Movements in Architecture

Singapore : Global Arts, 2006

Reviewed by Kim Williams

Keywords : Michael Ostwald, Baroque architecture, architecture of complexity, architecture of the fold

Via Cavour, 8
10123 Turin (Torino) Italy
kwb@kimwilliamsbooks.com

With the new architecture still in constant evolution, and its forms fragmented, bent, twisted and folded out of any immediately recognizable shape, a comparison with older, more canonical architecture also helps us to see the new, and to understand what we are seeing.

The aim of *The Architecture of the New Baroque* is to investigate the proposed claims regarding similarities between historical Baroque architecture of the 1600s (and its revival in the 1800s) to works of contemporary architecture which has been named "the new Baroque" (also known as the architecture of complexity or the architecture of the Fold). As author Michael Ostwald says early on, "Because the historic Baroque has been subjected to extensive historical scholarship it is used to provide a stable foundation for the remainder of the research" (p. 22). Comparing the new to the old is helpful because by revealing what the new is not, we are a step closer to understanding what it is. This kind of comparison works like a translation, allowing the reader (and the spectator) to grasp the meaning while appreciating the particular flavor or color of the original. What Ostwald does is examine the fidelity of the translations to discover if the touted similarities are superficial or profound (and thus if written theory is outstripping built projects). Along the way he examines not only architecture but the theoretical underpinnings, so that this small essay is unexpectedly rich. It allows those who have not followed the step-by-step unfolding of the movement to see it as a whole cloth (freezing it, of course, somewhat artificially at the point when the book was written, although the architecture itself continues to evolve). These days the rapidity and spread of theory means that its effect on design practice is almost immediate, much more so than when Guarino Guarini had a career of building experience behind him before writing his treatise on architecture.

The first step in Ostwald's study is a review of "historical Baroque." To accomplish this, the chapter is divided into several sections. The first describes the etymology of the term "Baroque". The second deals with the timeframe of the style. The third places the style in its cultural context. The final section examines the architectural theory.

The next chapter, "The Old and the New", begins correlating the key concepts of the twentieth century with those of the seventeenth. Historically, the dual cultural ideals of openness and systemization were expressed concretely through the architectural concepts of infinity and movement. Ostwald identifies four strategies used to evoke infinity: curvature and perspectival devices of the wall, changing proportional and formal relationships, blurring of boundaries, and naturalistic forms and decoration. The next four sections consider these in turn.

This is the heart of the argument: the questions of if and how the old and the new Baroque movements are related will be answered here. This brief description of how the argument develops already gives an idea of the care with which these problems have been addressed.

Ostwald makes it clear that comparisons between historic and new Baroque works take into consideration not only visual similarities but those of technique and intent as well. This reviewer appreciated hearing that, because while I found very little in the way of visual similarities, I believe the comparisons of technique and intent are convincing.

Even the descriptions of historic Baroque characteristics are often dependent on their contrast with those of the Renaissance: like the new Baroque, the old Baroque is best described by what it is not.

Ostwald's conclusions are guarded: if any artificial label can be justified, then that of New Baroque is justified, but in the final analysis the label itself is only relatively useful as a tool for analysis, criticism and study. Along the way, however, he does show us enough of the New Baroque to bring it into focus, which is in itself a good result. He cannot help but inject some criticisms along the way: Gehry's Experience Music Project in Seattle "veers dangerously close to cacophony, rather than melody" (p. 81); "Late Twentieth Century examples of scenographic architecture are relatively common but few are as carefully proportioned or controlled as the historic Baroque examples" (p. 77); "Gehry's illusion [in the Bilbao Guggenheim], unlike Bernini's [Scala Regia], is less seductive and more likely to evoke amusement ... than it is to suggest infinite movement" (p. 72).

But these criticisms of the architecture are not limited to the new, so critical balance is maintained. Historic Baroque also has its flaws: excessive emphasis on detail could prevent perception of a coherent whole; contemporaneous use of several of the strategies for achieving the desired aims could simply overwhelm rather than involve the spectator.

Ostwald also makes it clear that in spite of some arguable similarities, the rules of the game have simply changed. The palette of materials is much more varied today, for one thing. For another, today's design tool of choice – the computer – has altered virtually everything non-material: from the thought process of the designer, to visualization during design, to structure, to assembly.

This brings us to the final part of the book, which is a reflection on the implications of the New Baroque for architectural practice. Ostwald has addressed the implications and ethics of the use of computer-aided design at two Nexus conferences [Ostwald 2006, 2008]. His thoughtful consideration of the issue involved are a happy medium between unquestioning acceptance and reactive rejection.

The postscript of the book examines anthropomorphism and surface. The example illustrated, the 2001 Online Training Centre for Victoria University in Melbourne, reminded this reviewer of the facade explorations of Robert Venturi and Denise Scott Brown in the 1970s (like the Best Products building in Oxford Valley, Pennsylvania). Is this superficial play, or serious architecture? We'll need a few more years to know. But this present book is a thoughtful guide for keeping track of and following developments and placing them in a – if not the – context.

The book could have profited from a larger format for the sake of the figures, and careful editing would have smoothed out some of the rough passages, but these is minor censure of what is, overall, a model of timely, considerate architecture criticism.

References

OSTWALD, Michael. 2006. Geometric Transformations and the Ethics of the Curved Surface in Architecture. Pp. 77-92 in *Nexus VI: Architecture and Mathematics*, Sylvie Duvernoy and Orietta Pedemonte, eds. Turin: Kim Williams Books.

OSTWALD, Michael, Josephine VAUGHAN, Chris TUCKER. 2008. Characteristic Visual Complexity: Fractal Dimensions in the Architecture of Frank Lloyd Wright and Le Corbusier. Pp. 217-231 in *Nexus VII: Architecture and Mathematics*, Kim Williams, ed. Turin: Kim Williams Books.

About the reviewer

Kim Williams is editor-in-chief of the *Nexus Network Journal.*

Book Review

Giuseppe Dardanello, Susan Klaiber, Henry A. Millon (eds.)

Guarino Guarini

Torino: Umberto Allemandi & C., 2006

Reviewed by Kim Williams

Via Cavour, 8
10123 Turin (Torino) Italy
kwb@kimwilliamsbooks.com

Keywords : Guarino Guarini,
Baroque architecture

The long awaited, monumental volume dedicated to Guarino Guarini, edited by Giuseppe Dardanello, Susan Klaiber and Henry Millon is all it was hoped to be. Resulting from the 2002 seminar on Guarini held at the Centro Internazionale di Studi di Architettura Andrea Palladio in Vicenza, the present volume is the successor to the two-volume proceedings published in 1970 of the international conference that took place in Torino's Accademia delle Scienze in 1968. The list of scholars present at the 1968 conference was star-studded, including Rudolf Wittkower, Franco Borsi, Werner Mueller, Werner Oechslin, Manfredo Tafuri, Paolo Portoghesi, Henry Millon, Richard Pommer, Christian Norberg-Schulz and many others. The resulting collection of papers represented the latest word in Guarini scholarship up to that point. Now with the publication of this new book, Guarini scholarship – having in the meantime been enriched by the monographs of H. A. Meek (1988) and John Beldon Scott (2003) – is once again brought up to date, taking advantage not only of new information gleaned from the ongoing study of archives, etc. (and the dubious silver lining of the 1997 fire in the Chapel of the Holy Shroud, which made it possible to learn things about the Chapel which would otherwise have remained unknown), but also of new methods of investigation that have been developed over the intervening thirty-six years.

The volume is organized into five major sections: Part I, an introduction to Guarini; Part II, discussions of his intellectual and professional figure; Part III, an examination of his built and unbuilt works, which is actually two sections, the first a collection of photographs and reproductions of drawings and engravings, and the second, a substantial group of papers; Part IV, dedicated to the diffusion of Guarini's ideas throughout Italy to Europe and abroad. These major sections are followed by the scholarly apparatuses of bibliographies and a comprehensive index of names and places.

Of the scholars who contributed to the 1970 volume, only Henry Millon is present in the new volume to provide continuity.

The large format of this work makes it possible to accommodate a particularly rich graphic program. The central sections of splendid color photographs of Guarini's still extant works by Pino Dell'Aquila, together with reproductions of drawings and engravings compiled under the direction of Giuseppe Dardanello could almost form a monograph by itself. These images are of signal importance to understanding the various aspects of Guarini's architecture discussed in the text.

As far as the discussions of the more mathematical aspects of Guarini's architectural thinking is concerned, which will be of greatest interest to *Nexus Network Journal* readers, the 1970 volume includes treatments of Guarini and stereotomy by Werner Mueller, of Guarini and perspective by Corrado Maltese, and a very important discussion of Guarini's use of geometry by Henry Millon. However, the talk by Guglielmo De Angelis d'Ossat on proportional schemes and geometry did not result in a paper for the volume, and the laconic paper by Francesco Tricomi on Guarini as mathematician only whetted the appetite for more. Still, one of the fascinating aspects of Guarini's work is the prominence of mathematical – especially geometrical – ideas and even a paper such as Gianni Carlo Sciolla's on Guarini's treatise on fortifications opens the way to a study of the geometry inherent in it.

The papers in the new volume that deal with Guarini's mathematics are an important follow-up. Regarding theory, Marco Boetti's paper on the geometry of the vaults and interwoven arches provides an in-depth catalogue with dimensional and geometrical comparisons of eight of Guarini's dome structures in sacred buildings, while Edoardo Piccoli undertakes an examination of the vaulting in his civic architecture, placing the structures in the context of geometry and stereotomy. Aurora Scotti Tosini's discussion of the relationships between text and images in Guarini's written works makes evident parallels between illustrations in his *Euclidis Adauctis* and *Architettura Civile*. John Beldon Scott, discussing the changing tendencies that have characterized studies of Guarini's oeuvre provides a brief but instructive overview of studies on the geometric bases of Guarini's design process.

In the group of papers dealing with individual architectural projects, several papers are noteworthy for their attention to geometry. Giuseppe Dardanello's discusses Guarini's transformation of Castellamonte's oval plan for the Chapel of the Holy Shroud into a rotunda. Franco Rossi looks at Guarini and stereotomy. Andrew Morrogh's discussion of the plan of San Lorenzo uses both geometry and proportions as analytical tools. Henry Millon's examination of the church of the Immacolata Concezione in Torino tells how Guarini modeled space "not only with structures set up with circles, ovals and other curvilinear forms, but also schemes composed of triangles, squares, pentagons, hexagons, octagons and polygons in general." Gerd Schneider's analysis and reconstruction of Guarini's design for a church in Ceva includes geometric diagrams of the proportions of the facade, the major and minor axes, and the plan.

All papers in this book are impeccably documented. The comprehensive index of names and places – so often lacking in volumes such as this which are collections of papers – is a very important addition to rendering the book even more useful to scholars. The bibliography is ordered chronologically – good as a device for understanding the unfolding of Guarini studies, but less helpful as a tool for hunting down references than one ordered alphabetically (try finding the reference for Meek's monograph on Guarini if you are uncertain of the date of publication).

The innumerable references in all the papers of this present volume to those of the 1968 conference proceedings is confirmation of that collection's enduring value. This present book is destined to become equally valuable, and the editors are to be congratulated for having accomplished such an important task.

About the reviewer

Kim Williams is editor-in-chief of the *Nexus Network Journal.*

NEXUS NETWORK JOURNAL Architecture and Mathematics

Changes of Address: Please report any change of address at least 6 weeks before it is to take effect, giving your current as well as your future address.

Claims: Claims for issues not received should be made promptly – at the very last immediately upon receipt of the issue succeeding the missing number.

Cancellations: Cancellations must be received by September 30 prior to the new subscription year. No cancellation requests will be accepted for the current year once a subscription has been entered and payment has been processed.

Back Volumes of this journal are available. Prices are quoted at the current subscription rates per volume.

Correspondence concerning advertisements, general publishing matters and procedures should be addressed to:
Birkhäuser Verlag, Journals Marketing,
P.P. Box 133, CH-4010 Basel, Switzerland
Fax: ++41 61 205 07 91, email: nexus@birkhauser.ch

Abstracted/Indexed in
CompuMath Citation Index, Current Contents/Physical, Chemical and Earth Sciences, Journal Citation Reports-Science, Mathematical Reviews, MathSciNet, PASCAL, Science Citation Index Expanded, SCOPUS, Zentralblatt Math.

Visit our web site
http://www.birkhauser.ch

Printed on acid-free paper produced from chlorine-free pulp
© 2009 Kim Williams Books and Birkhäuser Verlag,
Basel • Boston • Berlin
ISSN 1522-4600 (electronic)
ISSN 1590-5896 (print)

Instructions for Authors

Authorship
Submission of a manuscript implies:
- that the work described has not been published before;
- that it is not under consideration for publication elsewhere;
- that its publication has been approved by all coauthors, if any, as well as by the responsible authorities at the institute where the work has been carried out;
- that, if and when the manuscript is accepted for publication, the authors agree to automatically transfer the copyright to the publisher; and
- that the manuscript will not be published elsewhere in any language without the consent of the copyright holder.

Exceptions of the above have to be discussed before the manuscript is processed. The manuscript should be written in English.

Submission of the Manuscript

Material should be sent to Kim Williams
via e-mail to: kwilliams@kimwilliamsbooks.com
or via regular mail to: Kim Williams Books, Via Cavour, 8, I-10123 Torino, Italy

Please include a cover sheet with name of author(s), title or profession (if applicable), physical address, e-mail address, abstract, and key word list.

Contributions will be accepted for consideration to the following sections in the journal: research articles, didactics, viewpoints, book reviews, conference and exhibits reports.

Final PDF files
Authors receive a pdf file of their contribution in its final form. Orders for additional printed reprints must be placed with the Publisher when returning the corrected proofs. Delayed reprint orders are treated as special orders, for which charges are appreciably higher. Reprints are not to be sold.

Articles will be freely accessible on our online platform SpringerLink two years after the year of publication.